This is an entertaining romp through th[e...]
telling the stories of key people to illustra[te...]
along the way.
– **Adrian Cockcroft**, former cloud architect at Netflix and AWS

A fascinating look back at the origins of our modern computing environment: how the technology evolved and, perhaps more importantly, how the people involved made it happen. As well as providing an informative history, Jamie Dobson draws out the lessons we can learn from how things developed and suggests how they might apply in the era of AI.
– **Steve Freeman**, principal engineer, Gousto

A roller coaster ride through the history of modern humanity and its technologies. This eye-opening story highlights the power of trial and error and how amazing progress can come through failure by learning from it.
– **Anne Currie**, founder of Strategically Green and author of *Building Green Software*

Jamie Dobson has pulled off the impossible – making the history of computing feel like a heist movie. From Edison's lightbulb moment to today's AI revolution, he shows how the best innovations happen when brilliant misfits are given the freedom to fail spectacularly.
– **Mark Coleman**, co-founder and chief evangelist at NetBox Labs

More than just a book on tech history, this is an enjoyable journey through the failure-fuelled cultures that powered innovation, ultimately leading to the cloud and AI. Thoroughly engaging and thought provoking.
– **Daniel Vaughan**, director of software engineering, Mastercard

Jamie Dobson weaves a captivating tale of humanity's tech evolution – from electrification to AI – with brilliance and clarity. A must-read for tech enthusiasts and novices alike, it's a captivating journey through innovation's past, present and provocative future.
– **Jan Wiersma**, technology executive and investor

A compelling exploration of the intertwined journeys of technology and humanity. With insightful storytelling, Jamie Dobson reveals how the pioneers of innovation have shaped not only the tools we use, but the very fabric of our society. This book is essential reading for anyone who wants to understand the profound impact of technology on our lives and the leaders who dared to dream.
– **Philippe Ensarguet**, VP software engineering, Orange

A must-read for anyone who wants to understand not just how technology evolves, but how great teams and bold ideas shape the future. If you care about innovation, leadership or the impact of technology on society, this book belongs on your shelf.
– **Jonny Williams**, chief digital advisor, Red Hat and author of *Delivery Management: Enabling Teams to Deliver Value*

A rollicking page-turner! Jamie Dobson's enthusiasm for the topic jumps through the page as he unpacks the extraordinary history of technology's evolution. From the electrification of society to the birth of cloud computing, this book is a whirlwind tour of the inventions, innovators and unintended consequences that shaped the world of cloud-native technology. Essential reading for anyone curious about how we got here, and where we're headed next.
– **Ian Miell**, partner at Container Solutions

Technology's revolutionary journey, artfully rendered.
– **Simon Wardley**, creator of Wardley Maps

An insightful journey through the history of cloud and AI technologies as well as the people who made things happen.
– **Chris Swan**, co-host, Tech Debt Burndown podcast

Visionaries, Rebels and Machines

The story of humanity's extraordinary journey from electrification to cloudification

Hope you enjoy the book, Glynn.

All the best, J.D.

JAMIE DOBSON

Visionaries, Rebels and Machines

ISBN 978-1-915483-15-7 (paperback)

ISBN 978-1-917490-07-8 (hardback)

eISBN 978-1-915483-16-4

audio ISBN 978-1-917490-06-1

Published in 2025 by Right Book Press

Printed in the UK in May 2025

Manufactured by
Sue Richardson Associates Ltd.
Studio 6,
9, Marsh Street
Bristol BS1 4AA
info@therightbookcompany.com

EU Safety Representative
eucomply OÜ
Parnu mnt 139b-14
11317 Tallinn
Estonia
hello@eucompliancepartner.com
+33 756 90241

© Jamie Dobson

The right of Jamie Dobson to be identified as the author of this work has been asserted in accordance with the Copyright, Designs and Patents Act 1988.

A CIP record of this book is available from the British Library.

All rights reserved. No part of this book may be reproduced, stored in a retrieval system, or transmitted in any form or by any means, electronic, mechanical, photocopying, recording or otherwise, without the prior written permission of the copyright holder.

For Brendan, who waited with me in the rain.

Contents

Preface	ix
Introduction	1
Prologue	9

Part 1: From the Enlightenment to enlightenment

1. A digital and binary system of communication	19
2. The right call	27
3. The invention factory	33

Part 2: Electrons and nuclear bombs

4. The peculiar birth of a peculiar company	47
5. The computer	55
6. The wasp and the fig tree	65

Part 3: Those who knew the score

7. The Alto and the internet	85
8. The rise of utility computing	97
9. The dramatic fall of utility computing	109

Part 4: Cloudification

10. The magical computer store	127
11. Saturday night at the movies	139
12. A womb with a view	153
13. The computer as a community	163

14. Talent density	173
15. The computer and the company	183
16. The secret life of teams	195

Part 5: The enchanted loom

17. The end of the world (of work) as we know it	215
18. Only the paranoid will survive the coming wave	229
19. Unintended consequences	245
Epilogue	267
Afterword: Galvani's frog	285
Acknowledgements	289
Bibliography	297

Preface

The story of cloud computing is a rip-roaring yarn that spans not just the past few decades but the whole of human history. Among the people it includes are a distant relative of Sigmund Freud who thought it was a good idea to rent out videos over the web and a man who had an idea as crazy as renting videos online: selling books online. In making his web shop work, the bookseller invented a way of organising and renting out computers.

The story includes engineers and scientists passing electricity through anything and everything, including frogs, dogs and, in awful experiments, people. Later, silicon, that reluctant conductor of electricity, got the same treatment. That led to integrated circuits. We call integrated circuits microchips. They make computers work.

The story of the cloud contains at least three premature deaths and billions of births, including my own in 1976. The timing of my arrival allowed me to witness first hand, and shape ever so slightly, the trajectory of cloud computing.

The story includes scientists and engineers who, after fleeing Hitler's Germany, went on to make stunning contributions to the fields of psychology and computing, which in the early years were often indistinct from each other as avenues of research.

The story of the cloud also includes a snot-nosed teenager who stole computer cycles from a computer company in what was one of the world's first cybercrimes.

Jules Verne makes an appearance, too. His works of fiction

foreshadowed the electrification of our societies. It was this process that would one day allow us to cloudify those societies. Once we understand electrification, we are well on the way to understanding cloudification.

Why am I telling this story and why am I telling it in this way?

It all started when a young colleague asked me, 'What is the cloud?' I could answer, but not quickly. After noodling on the question one summer, I realised that the only way to answer it was to explain the cloud in its historical context. This not only tells us where it came from but helps us to understand where it might take us. As I poked around, it turned out that the cloud and computing's rich history was joyous, exhilarating, scandalous, whimsical, sad, at times hilarious and, in its latest chapters, portentous.

Artificial intelligence is a tidal wave that is currently breaking, and breaking things, all around us. Because it's already here, commentary on artificial intelligence is a peculiar, often confusing and almost always out-of-date mixture of description and prediction. There's some consensus on the past but not much consensus on the future. In the final few chapters of this book, I will try to bring us one step closer to a consensus – although I accept that I won't succeed. Right now, nobody really knows how the future will unfold. It is in those chapters and especially the final one that I switch from computer nerd to (amateur) journalist. It is, after all, the job of journalists to write history's first drafts.

You, the reader

I started this book with one target reader in mind. His name was Sam. Sam is by now a figment of my imagination but once upon a time he was real. He was the young colleague at Container Solutions who asked me that fateful question, 'What is the cloud?' As I tried to answer that question and as the book took shape, three different groups of readers, each with one thing in common, emerged. The first group crossed my path late in the process. The organisers of the

London DevOps Meetup, a monthly community event, wanted to know if I could speak one evening. Since I was right in the middle of writing Chapter 10, I told them I could tell the story of where the cloud came from. But I could only do that after spending a few minutes on the personal computer, the internet and something called Moore's law. Fine, they said.

To my surprise, after the talk, a gaggle of recent graduates surrounded me. Starved of history, they wanted to know more about the olden days. The neural networks of my mind exploded into life as I prepared to answer questions about vacuum tubes and time-sharing. They quickly stopped me and clarified what they meant. Their questions were not about the 1960s. They were, instead, about the 1990s. Since they were born in the Noughties, to them the 1990s were the olden days.

The second group contains curious and intelligent (but not necessarily educated) professionals. Not long after the DevOps meetup, friends of mine came over for drinks. One of them works in film and television as a lighting engineer. His skills and dexterity will not be replaced by a machine any time soon. Two are computer programmers. They still had jobs but with cuts looming, as companies learn to do more with less, both live in fear that one day soon they will be unemployed. The fourth is a wealth manager specialising in artificial intelligence. She shared a dirty little secret with me. Her team cannot value businesses in the artificial intelligence space, just like wealth managers once couldn't value web or cloud businesses. She told me her team, and her peers across the industry, were too embarrassed to admit that they had no idea what they were talking about. The world, she told me, is crying out for an explanation as to what artificial intelligence is, what it can do and what it might mean for society.

I told my friend that I had some good news for her. Artificial intelligence is the cherry on top of the cloud computing cake. You can pick the cherry, which is what my friend and her team had already done, but it will be unsatisfying. If you want to enjoy the cherry, I said, you have to eat it with a mouthful of cake. This book, I told her, is that mouthful of cake.

The third and final group is full of people who lead their businesses and have the daunting task of innovating, but aren't quite sure where to start. This group may have people in it who have tried to use the cloud to outpace their competitors but may be confused by the results of their fledgling attempts. For these readers, this book is an air-conditioned tour bus and I am the guide at the front with the microphone. Cake will be served throughout.

Sam and everybody in these three groups have something in common. They are practice rich but theory poor. In my experience, a sliver of theory, not to mention a dollop of history, helps the medicine go down. This book is that sliver of theory and that dollop of history. I hope you enjoy reading it as much as I enjoyed writing it.

Jamie, London, September 2024

Introduction

The short history of a utopian technique

In the last century, two titanic ideas collided with each other. The first was humanistic management; the second was digital computers. Humanistic management is about creating the conditions for people to grow through failure. Likewise, through a trial-and-error process of discovery, digital computers let programmers do the same.

Organisations built around humanistic management techniques outpace their competitors because they have learned to learn from failure. The cloud, like its antecedents, time sharing and personal computers, supercharges this process. If psychologist Abraham Maslow were alive today, he would say the combination of cloud computing and humanistic management is a utopian technique. He would be right.

The relentless march

Humanity's inexorable journey towards this utopian technique accelerated in the magical 1880s. In that decade, modern Prometheans raided Mount Olympus and brought back with them Zeus's thunderbolts, which they then caged in a grid of restless energy (Schewe 2007). Once there, with the flip of a switch, these thunderbolts could be brought back as power and lights for factories and houses.

Since humanity had already worked out how to send bursts

of electricity through telegraph cables that ran alongside railway lines, once the grid arrived, electricity was about both power and communication. Less than 100 years after Thomas Edison invented the light bulb, those poles and those bursts evolved into the internet.

The rapid evolution of telecommunication technology, like the other children of electricity such as the microchip and digital computer, was matched in the last century by humanity's deepening understanding of human nature.

In the early decades of the last century, behaviourism brought scientific rigour to the field of psychology. By doing that, it displaced the work of Sigmund Freud, who considered all of human behaviour to be driven by unconscious urges. Later, though, behaviourism was superseded by humanistic psychology. The humanistic psychologists saw human nature and behaviour in holistic terms, something more than stimulus and response. They cared about creativity, love, belonging and becoming. In other words, they cared about what makes us human.

Then, quite by accident, humanistic psychology morphed into humanistic management. A gigantic leap for Abraham Maslow, and therefore humankind, took place in California, where a company organised themselves around humanistic principles. It was during his time studying Non-Linear Systems that Maslow realised that the proper management of work helps people to fulfil their potential. That led Maslow to a stunning conclusion: management was nothing short of a 'utopian or revolutionary technique' (Maslow 1998). I'll return to this in more detail in Chapter 15.

About ten years later, in the 1970s, the progress humanity made with computer technology met Maslow's conclusion at a company called Intel. At that time, Intel was a small semiconductor firm, but not long after, due to the rise in popularity of the personal computer, it became one of the most important companies in the world.

The men responsible for Intel's rise were its founders, Robert Noyce and Gordon Moore, and its first employee, Andy Grove. As young scientists, all three shaped the development of semiconductor technologies. However, by the end of the 1970s, they were well and truly entering into middle age and doing so as executives. That's why they no longer had to only solve technical problems. They had to solve people problems, too.

In his role as chief operating officer, Grove set his sights on a gnarly people problem. He wanted to motivate some of his employees while helping those who were already motivated to focus. Armed with what he knew about technology, learning from failure and, importantly, his own understanding of Maslow's work, Grove performed an act of mental ju-jitsu that forever changed how knowledge workers were managed. He created a 'hack' based on an old idea called management by objectives (MBO) and the then newish ideas of Maslow. This hack was called 'objectives and key results' (OKRs).

OKRs were a system, Grove claimed, that created focus 'par excellence'. His system, however, was only partially about performance. Grove thought it essential that the objectives stretched his employees so that, about 50 per cent of the time, they failed. It was through these failures and the lessons they taught that Grove's people, and therefore Intel, grew.

With OKRs, Grove had taken a step towards implementing Maslow's utopian technique. Such was Grove's influence, from that point onwards humanistic management and the development of technology would be forever wedded. This marriage, dating all the way back to the 1970s, is the practical and philosophical core of cloud computing. It is also the central theme of this book.

My excellent adventure

I arrived right in the middle of all this, in the summer of 1976. There was a heatwave at the time that my mum and her generation have not stopped talking about. They talk about it as if it was our fault.

Unbeknown to my parents, as Grove fiddled with OKRs in California, and as they tiptoed down the streets to avoid trampling the plague of ladybirds that descended on them that summer, my fate and that of thousands of children were being twisted by a computer scientist called Christopher Evans.

Evans contemplated the seismic effects the computer was having and would have on society. A few years later, in 1979, he published a book, *The Mighty Micro: The Impact of the Computer Revolution*. In the same year, Evans wrote and then presented a documentary that was broadcast in the UK by ITV (Independent Television), a free-to-air channel that was created to compete with the state-run BBC (British Broadcasting Company). Evans did not live to see its wild success or the impact it had on the psyche of the British. He died of cancer before it was broadcast. He was only 48.

Finding itself playing catch-up, in the early 1980s the BBC started the BBC Computer Literacy Project. This was, however, more than a series of documentaries. To supplement the shows, and sparing no expense, the BBC commissioned a personal microcomputer for use in schools. The computer was built by a company called Acorn but was known as the BBC Micro to most or, by me and my mates, as the BBC. It was through this chain of events that a computer arrived in the corridor outside our music room at St Anne's junior high school when I was nine.

My first program

The BBC shipped with a programming language called BASIC. Because of his dad, my friend Kristian knew some of BASIC's commands. Kristian helped me, like thousands before and millions after, write my version of a beginner's program.

```
10 PRINT "JAMIE"
20 GOTO 10
```

Line 10 tells the computer to print 'JAMIE'. Line 20 says go back to line 10 and that plunges the program into an 'infinite loop'. I had taken the first step of an excellent adventure. Not long after, I would take another.

Summer rolled around. I was on the verge of learning the secret to becoming a great programmer: the studying and copying of other people's code. Greenwood Avenue Library wasn't far from our house. The librarian there let me take books from the adult section. That was forbidden. But I had wanted to read Frank Herbert's *Dune,* and she set a precedent when she let me take it out. On this particular trip, I found a book full of BASIC programs. The librarian let me take that out, too.

Back home, on a second-hand Commodore 64 that my mum had bought, I meticulously copied a program that, when it ran, made two coloured squares flash on the screen. When the two squares were the same colour, the player (me) hit the space bar. If they did this before the colours changed again, they got some points.

I couldn't play the game because the computer was hooked up to an old black-and-white television. I couldn't save the game, either, because I didn't have any blank tapes. Neither of these things mattered. Playing or saving the game were a million miles from the point. The point was that I'd made a computer do my bidding. The ramifications were obvious. I switched the computer off and as the charge in the chips died, so too did my program. What didn't die was the idea of what can be done with computers. I went to bed that night, and almost every night since, dreaming about what can be done with computers.

A brave new world

I'd stepped into a new world. In this world, knights, in a computer game called *Joust*, rode on giant chickens. In this world, I wrote computer programs iteratively, uncovering their form like a sculptor uncovers statues hidden in the clay. In this world, I was bound by the immovable laws of physics and the unbendable rules of the computer. Yet, in this world, I was unbound by the simple fact that if it can be simulated, it can be programmed into a computer. The only limit was, therefore, the imagination, which explains why in this world the knights rode chickens.

I now know it wasn't only computers that I came into contact with that summer. I'd come into contact with creativity, too. A puny ten-year-old and even punier microcomputer had together done something miraculous that neither could have done alone. A secret had been discovered and a lesson had been learned. The lesson was that computers and creativity are two sides of the same coin, the yin to each other's yang. The secret was… well, we'll get to that.

This book

The computer arose from systems of humanistic management. Once invented, it was then used to supercharge the creative process, which means it was used to supercharge its own development and that of the programs which ran on it. That's how the human race ended up with *Joust*, Microsoft Word, the web, Amazon.com and the cloud within the flutter of an evolutionary eye.

Throughout the last century, computer technology and humanistic management developed in parallel. Sometimes they came together explicitly, like when Grove used Maslow's work as the foundation of OKRs. Sometimes they came together accidentally, like when a man called Bob Taylor created the perfect environment for his team of researchers to invent the personal computer. However, there's a problem when trying to understand the relationship between computing and humanistic management.

Technology is irresistible to historians, science fiction writers and managers who should know better. This explains why, when you ask people what Edison invented at Menlo Park, they say the light bulb. A better answer is that, at Menlo Park, Edison invented a system of management that foreshadowed the great research organisations of the 20th century and, to go with it, a culture in which people were not afraid to fail.

It's impossible to understand the cloud, and how to succeed with it, without understanding computers, the systems they're made from, the types of engineers and scientists who built them and the systems of management that made them all feel safe enough to fail. That's what this book is about.

We're now moments away from setting off on a romp that explains the unbreakable link between computers, creativity and management. We're going to do that in five parts.

- In Part 1, **From the Enlightenment to enlightenment**, we will, for the first time, cross paths with systems, those who build them and the types of management in which they flourish.
- In Part 2, **Electrons and nuclear bombs**, we'll learn about the birth of the digital computer and how, from within the bowels of organisations that altered the trajectory of how knowledge workers should be managed, it was combined with telecommunications technology to stop humankind from destroying itself.
- In Part 3, **Those who knew the score**, we'll see that the history of technology is all about the marriage of ideas. That process went into overdrive in the 1960s and 1970s. The sexual revolution of those decades was mirrored by marriages of ideas in computing, the consummation of which gave us the personal computer, the microprocessor and, 40 years ahead of schedule, cloud computing.
- In Part 4, **Cloudification**, we'll see that by the 1990s all the pieces for the wide-scale cloudification of society were in place. Soon after, whole industries were upended but so too was humanity's understanding of management when, in 2014, a mystery more than 100 years in the making was finally solved.
- In Part 5, **The enchanted loom**, we'll see that improvements in cloud computing and theoretical leaps in a technology called machine learning led to what was hitherto thought impossible: machines that can think.

Where to now?

The story of the cloud is the story of humanity and therefore, whether you know it or not, is the story of you. We're about to go on a great adventure, the contours of which collide with the very limits of what it means to be human. That collision begins in the humblest of places. It begins in a bookstore.

Prologue

In the beginning was the word, and the word was Jeff's

It's hard to pinpoint the exact moment that Jeff Bezos made it. It was not, knowing what we know about the turbulent years that followed, his appearance on *Time* magazine's cover, in 1999, as its Person of the Year.

What about 2019 when, along with Lin-Manuel Miranda, the creator of the musical *Hamilton*, Jeff's likeness was added to the Smithsonian's National Portrait Gallery in Washington, DC? Michelle Obama, Hillary Clinton and Smithy from *Gavin and Stacey* (aka James Corden) danced in the crowd as Earth, Wind and Fire played on stage (Stone 2021). Surely that was the moment? Then again, maybe not. He had, after all, not yet been to space. That happened in 2021 when, on 20 July, he blasted off in a rocket that looked an awful lot like Dr Evil's from *Austin Powers: International Man of Mystery* (Reuter 2021).

How did this happen? How did a man who less than 30 years earlier started a mail order bookstore get catapulted into the Smithsonian and not long after that into space? How did this man acquire so much wealth and power that critics and internet trolls alike, including the president of the United States, felt compelled to take him down? Amazon had, after all, only just survived the dot-com crash of 2002, was written off as 'Amazon.toast' and then on the other side of the crash met financial analysts who said that if it walked and talked like a mail order bookstore, then *it was a mail order bookstore* (and should be valued like one) (Bezos 2021; Stone 2021).

What remarkable transformation took place in the years between the dot-com crash and that space flight in 2021? How had Amazon moved from behind the wheelie bins, where books were left for their customers, and into millions of kitchens where users installed a voice activated, *Star Trek*-like artificially intelligent personal assistant called Alexa? How had Amazon Prime mutated from a convenient, flat-fee system to get around the bugaboo of mail order shopping, the hassle and opaqueness of postage and packaging costs, to an on-demand movie and TV streaming service where some of its programmes came from Amazon's own film studio? (Gordon 2016; Stone 2021).

A culture of creativity

The answer to these questions lies in Amazon's culture of innovation.

Amazon's journey started simply enough, all the way back in 1994, when an emerging home appliance, the personal computer, had a new application written for it, the web browser. Web browsers read files from remote computers in far off and sometimes mysterious places using a communications system called the internet. Bezos thought that the computer, the browser and the internet could be used to commercialise the thing they were cobbled together to create: the World Wide Web. He would do his bit by selling the humblest of products: books (Bezos 2021; Stone 2014).

At that time, Bezos was 30. Compared to David Packard and Bill Hewlett and the Steves – Wozniak and Jobs – that was old to be setting off on an entrepreneurial journey. Packard and Hewlett were well under 30 when they started Hewlett-Packard. Their big break came when they sold eight audio oscillators to Walt Disney, who needed them for his 1940 movie *Fantasia* (Waldrop 2018).

The Steves, after helping to design and build the arcade game *Breakout*, the same *Breakout* that only 37 years later, because of advances in cloud computing, was mastered by an artificial intelligence, also caught their break when they were well under 30. That happened when they started a personal computer company. They called it Apple.

Like Hewlett-Packard and Apple, Amazon started in its founder's garage. The similarities, however, end there. Jeff Bezos had not started a technology company. He had started a mail order bookstore. Through an online catalogue that reached all of America, Amazon would sell books. This meant it had a lot more in common with the catalogue business that Richard Sears and Alvah Roebuck had started in 1893 than it did with Apple or Hewlett-Packard (Gordon 2016).

A good bet

As 1995 rolled around, the commercialisation of the fledgling web started to look like a good bet. That year, Microsoft caused a sensation, and miles and miles of queues, when it released its internet-enabled operating system, Windows 95, which (of course) came in a cellophane-wrapped box full of disks and a user manual.

In that same year, as more users got online, some with the help of Windows 95, orders for Amazon's books rocketed. Some of those buying books did so for the fun of it, like people used to get electronic calculators out at parties in the 1970s. Others, the dedicated early adopters and believers in these new technologies, ordered to prove they worked. But book lovers did most of the shopping because they had a way to browse, obligation free, a seemingly endless array of titles.

Not long after this initial burst of success and armed with only a sliver of personal information, Amazon's algorithms, way before artificial intelligence joined the party, made better recommendations than a bookstore clerk ever could or, given how shy most of them are, ever would. The system, in other words, worked. The web browser, pointed at Amazon.com, gave customers unfettered access to a selection of popular and specialised books from the comfort of their own homes. It was a book lover's dream come true.

One-click arrived two years later, in 1997, in the same year that Amazon landed on the stock exchange with an initial public offering of its shares. Two after that, in 1999, Bezos's mug graced the front cover of *Time*. The secret to this remarkable rise was Amazon's culture of creativity. In practice, that meant fast decision making, constant invention, a willingness to adopt new and emerging technologies and, of course, a willingness to fail (Stone 2021). The same characteristics appeared in Thomas Edison's 'invention factory', showing that technologies may have changed in the last century but the process of innovation had not.

Chaos

Like all businesses that allowed their customers to shop online, Amazon needed lots of computers, lots of data storage and lots of networking equipment. Collectively, this is called information technology (IT) infrastructure. A simple website can be hosted on a single computer but a web-scale application such as Amazon.com, with millions of customers, cannot. This is why, at the turn of the century, companies with web-scale applications had to learn how to manage their IT infrastructure. If they did not, their computers or the software running on them would fail. When that happened, their business would start to fail, too.

The chaos behind the scenes at Amazon, especially in IT, was jeopardising its operations and grinding its ability to innovate to a halt.

A cloud is born

The speed, or lack of it, with which IT infrastructure can be provisioned is called the *IT infrastructure bottleneck*. This drove those who built web-scale applications mad. Amazon was the only company to seriously do something about it. Why them? Well, out of sight and therefore out of the minds of Wall Street's bean counters,

Amazon focused on the long term. It valued creativity, which the IT infrastructure bottleneck was getting in the way of, wanted to move quickly and not break anything because it could not afford to, and was adamant that it should, could and indeed would have a system that made experimentation and therefore invention easy (Bezos 2021). This explains why, during those years when the IT infrastructure bottleneck was strangling the life out of Amazon's creativity, Bezos screamed out 'developers are alchemists and our job is to do everything we can to get them to do their alchemy' (Stone 2014).

By solving the IT infrastructure bottleneck, Amazon created a system that allowed its own developers to provision computers with as little fuss as its actual customers already ordered books. Its alchemists were, in other words, finally free to do their alchemy. It then dawned on Amazon, first slowly and then epiphanically, that the system it had created would let *all* the world's alchemists do their alchemy…

Amazon solved its IT infrastructure problems, which were expensive and dictated the pace of innovation. In doing so, it optimised its low-margin business. By choosing to help other companies optimise theirs, Amazon's journey to becoming a genuine technology company began.

This is how, in the mid-Noughties, Amazon found itself teetering on the brink of performing something akin to a miracle. That miracle was, of course, cloud computing. Its remarkable impact propelled Bezos to the Smithsonian and its stratospheric profits shot him into space.

The cloud as a 'grid' of computer 'power'

What had Amazon actually invented? Nobody really knew the answer to that question. The cloud was hard to understand and hard to explain, even for those who had invented it. This is why Bezos, in 2009, reached for an analogy. He said that, once upon a time, when the electricity grid came online, factories jettisoned their power generators. The cloud, he said, was a bit like that but for computers. Companies building web-scale applications could plug into this 'grid' of computer 'power'. In doing so, they took advantage of the world's most successful online retailer's computers just as factories had once taken advantage of the grid (Rose 2009). If that were true, Bezos was to computers what Edison once was to electricity.

Was this grid analogy accurate? It was, partly. Amazon spent a lot of money on building its cloud, just as Thomas Edison had once spent a lot of money building that vast machine of restless energy called the grid (Schewe 2007). Businesses dabbling with web-scale applications plugged into a vast 'cloud' of computing infrastructure just as they had plugged into the vast infrastructure of the electricity grid.

In other ways, though, it was misleading. The grid analogy said nothing of fast decision making, reinvestment or innovation. The cloud as an 'engine' of experimentation was nowhere to be found. There was no mention, in other words, that Amazon took failure seriously. The grid analogy went on to cause untold misery for those who conflated the reliable and regulated electricity grid with Amazon's so-called 'grid' of computing 'power'.

How and why? The reason, or the secret I promised to share in the introduction, which I discovered that summer with my Commodore 64, was that there are not two things involved in computer programming, the programmer and the computer. There are in fact three. One is the computer. The other is the programmer. The third, however, is the *process* by which the programmer *interacts* with the computer. Companies and their programmers failed with Amazon's cloud because it wasn't enough for them to just use their computers. They had to instead use them in the same failure-fuelled, experimental, interactive way for which they had been designed.

Earned luck

Of course, the misery of some of Amazon Web Services' early customers doesn't change the fact that Amazon invented cloud computing as we know it today. It was a remarkable and unexpected achievement. It took a further seven years for any competition to arrive (Bezos 2021). Their timing was perfect and the luck they made for themselves, impeccable. Amazon, the technology company, was off and running.

Where to now?

If we are to understand the most common and most misunderstood analogy used to explain the cloud, then we have to understand both the story of the grid and the 'invention factory' that gave birth to it.

That story may end in 2025 with artificial intelligence but it begins in the 18th century. A domino toppled on an Italian's workbench set in motion a chain reaction that ended with artificial intelligence but started with the twitches of a frog's legs. It is to that workbench and those legs that we now turn.

Part 1

FROM THE ENLIGHTENMENT TO ENLIGHTENMENT

Chapter 1

A digital and binary system of communication

As the 1700s and along with them the Enlightenment drew to a close, and decades before Thomas Edison actually enlightened society, humankind captured a few sparks from Zeus's thunderbolts and bent them to its whims in a rudimentary but practical system of communication called the telegraph (Landes 2003). The telegraph worked because an Italian scientist made a curious observation. That observation brought forth an invention and boosted the field of neurology, but cost him a friendship.

The voltaic pile

Alessandro Volta invented the battery in 1799 while trying to win an argument with his friend, Luigi Galvani, about why, when a clip touched a dead frog's leg, it twitched. After observing it, Galvani thought it had something to do with electricity that was intrinsic to biological tissues. He hypothesised that electricity in the frog's leg escaped when it formed a circuit. The release of this electricity caused the legs muscles to contract (McComas 2011). Galvani called this animal electricity.

Volta thought the opposite. Because Galvani had laid the frog's leg down on a copper plate and touched it with a zinc clip, Volta hypothesised that electricity within the metals, when they formed a circuit with the leg, caused the twitch. Additionally, Volta suspected the bimetal nature of the circuit played a role. He confirmed this by placing two different pieces of metal on his tongue, which made it tingle (Jorgensen 2021).

Volta further confirmed electricity's role in the body when he created a single circuit that included his own eye, his tongue again and a frog's leg. When he closed the circuit, the leg twitched, his tongue tingled and light appeared in his eye (Jorgensen 2021). Volta therefore discovered that electricity played a role not just in movement but the senses, too.

Volta then layered silver and zinc discs on top of each other in a brine-soaked solution (McComas 2011). He connected the top disc to the bottom disc to form a circuit. Electricity flowed without the involvement of any animals (or their legs).

Volta may have felt smug about what at first sight looked like a physics experiment and conclusive proof he was right. Like Galvani, though, he was also partly wrong. A chemical reaction occurred between the two metals that generated electricity and, incidentally, corroded them both. In a nod to his former friend, Volta called the generation of an electric current through chemical reaction 'galvanism'. 'Galvanic corrosion', on the other hand, is what happens to the metals during the reaction.

Because the whole argument had started with electric eels and how they create electricity, Volta called his contraption an artificial electric organ. Because the discs piled up on each other, his contemporaries called it a voltaic pile. The piles, or 'cells' in the electric 'organ', when stacked together looked like a *battery* of cannons firing at the same time (Jorgensen 2021).

A solution looking for a problem

Cheap, easy to assemble and scale, because extra discs increase the charge, plus able to sustain a consistent voltage, batteries contained the potential to be put to practical uses. What uses? Nobody knew. The battery was a solution looking for a problem.

The telegraph

It eventually found one in the telegraph. The telegraph included a key for sending a message; in later versions, a sounder on the other side; signal re-transmitters, known as relays, which amplified the signal; and Volta's batteries to power the line. The telegraph was, in other words, a simple system, explainable to intelligent people.

Resistance

To William Fothergill Cooke and Charles Wheatstone in England and Samuel Morse over in the United States, it must have seemed that there were few intelligent people around.

On both sides of the Atlantic, the telegraph bumped into scepticism. Visible, and therefore understandable, optical telegraphs' rearrangeable arms spelled out messages. The original electrical telegraph, on the other hand, communed messages to its priest-like operator through wobbling needles.

The incomprehension about the telegraph explains why members of Congress refused to spend taxpayers' money testing it. The representatives also knew about Victorian parlour tricks, like the kiss of the electrified (and usually beautiful) young woman, and thus suspected the telegraph to be another con job.

From our vantage point in 2025, it's difficult to imagine how our ancestors didn't recognise the transformative nature of the telegraph. There was a reason for this, however – and it's not because they were daft.

The crossover decades

Upon their arrival, epoch-changing technologies often coexist for a few decades with an existing era's attitudes and understanding. Within those crossover decades, the technology's promise must break through the noise of the day-to-day chores of everyday people while capturing the imagination of politicians who are caught up in the challenges of the day. The crossover decades explain why new technologies are enthusiastically welcomed (by those who studied or used them) and mistrusted or ignored (by almost everybody else) (Smil 2005).

In the crossover decades, the telegraph stirred up a normal, human and therefore predictable reaction. This explains why members of Congress would not prioritise a meeting with Samuel Morse over dealing with the tensions that were at that time mutating into the causes of the American Civil War.

Similarly, it's hard to imagine housewives on either side of the Atlantic getting excited about the telegraph when their never-ending chores at home – fetching water from wells, bringing coal in for the fire and scraping its remnants out in the morning – took up almost all their time and energy (Gordon 2016).

Everyday people saw new technologies such as electricity as curiosities or, if too far removed from their day-to-day existence, bonkers (Smil 2005). This led to Morse and Cooke and Wheatstone learning a lesson that all technologists eventually learn: until they see something working within the context of their day-to-day lives, people rarely understand it.

The demonstration

The only practical way to overcome the dollops of scepticism on both sides of the Atlantic was with a real-world demonstration. In England, that happened when Cooke and Wheatstone, in desperation after rejection by the military, convinced the Great Western Railway to connect London's Paddington Station with Slough, in Berkshire (Standage 1998).

Like the military, Great Western doubted that any good would come of it. But the telegraph and the railways went together like frogs' legs and zinc clips. In those days, stationmasters coordinated trains with error-prone signals and even more error-prone printed timetables, which broke down whenever the trains did. With the telegraph, train stations communicated with each other in real time about arrivals, departures, delays and cancellations, which they in turn announced to their passengers.

However, despite the speed and reliability of the telegraph, as the experiment progressed, doubts still lingered in the minds of the railway executives. Then the telegraph received a public relations boon when it was used in a way that nobody, including its inventors, saw coming.

Unimagined applications

On 6 August 1844, news of the birth of Queen Victoria's second son, Alfred, born at Windsor Castle, just outside Slough, electrified the wire back to London. Within 40 minutes, *The Times* hit the streets with the announcement (Standage 1998). Once the Duke of Wellington heard, he made his way out to Windsor for the celebration dinner but forgot his dinner jacket. He telegraphed back to London and the jacket made its way to Slough on the next train.

In a further, but yet again unforeseen, demonstration of its usefulness, the telegraph turned crimefighter against Fiddler Dick (Standage 1998). Rather than a distant member of the Royal Family popping to Slough or their resident violinist, Dick was a pickpocket extraordinaire who relieved passers-by of their possessions before escaping on a train. Once the telegraph arrived, the crime and the criminal's description click-clacked their way from the telegraph key to the sounder at the other end. That was the end of Fiddler Dick's criminal antics.

Just as the telegraph was receding from the public's imagination, a much more sinister case catapulted it back into the newspapers. One year after the birth of Prince Albert, John Tawell murdered his girlfriend in Slough and escaped to London on the train. Because the telegraphic alphabet missed a 'Q', the operator's message to London, describing Tawell's dark overcoat, said that he 'is in the garb of a kwaker'. The police kwickly arrested Tawell. After his execution, the newspaper reported that the telegraph wires were 'the cords that hung John Tawell (Standage 1998).

The barriers between time and space

Over in the United States, Morse convinced the US Congress (but only just) to help him prove the concept of the telegraph by linking Washington, DC to Baltimore. It went wrong for Morse when the wires he used rusted, which stopped them conducting. (Galvanisation, the process of protecting metals from galvanic corrosion, was not yet in widespread use.) In a panic, and only as a short-term measure as he proved the idea, Morse suspended the wires above the ground on big, wooden poles. News soon flowed between the two cities. Resistance then shifted its shape. Baltimore's religious leaders said the telegraph was black magic. However, it soon convinced more sceptical minds, including the religious ones, slowly at first and then quickly. Later, the *Baltimore Sun* reported that 'time and space has been completely annihilated' (Rosen 2012).

Not surprisingly, the telegraph went viral. Poles and wires soon ensnared the whole planet and Morse learned that quick 'hacks' to a system in its development can lead to unintended consequences that are hard or even impossible to reverse later. The telegraph annihilated time, space and, with its poles, the world's landscapes.

Dots and dashes

The final piece of the telegraphic puzzle was a universal code that allowed operators, often in different parts of the world and therefore fluent in different languages, to communicate with each other. Morse's code mapped the letters of the alphabet to unique sequences of dots and dashes, which the telegraph key converted into an electrical signal. Since the code consisted of two states, the telegraph was a binary system of communication, arriving 100 years before the boffins at Bell Labs invented information theory.

Where to now?

Along with steam engines and steamships, the telegraph annihilated the barriers between space and time. Humanity, though, was just warming up. Inventors laid, in what Smil calls 'the miraculous 1880s', the foundations of modern life. These foundations included the electricity grid, the internal combustion engine, wireless communication and the telephone, which started life off as the harmonic telegraph and, as we're about to find out, is shrouded, even to this day, in controversy. We're off to that miraculous decade now to meet that Scottish Victorian furball of indecision, Alexander Graham Bell.

Chapter 2

The right call

Alexander Graham Bell filed what has been called the most valuable patent in history on Valentine's Day in 1876 (Gordon 2016). However, the key component that changed sound into an electrical signal was missing from Bell's patent application. He hadn't quite figured that bit out. This was partly because he had been messing about with his dad's Visible Speech system (to help deaf people learn to speak) but mainly because he was a bit useless.

Rather less useless was a man called Elisha Gray, who worked for Western Union, the telegraph company. His patent application had a detailed drawing of a device that would change sound into an electrical signal. Gray submitted his patent just a few hours after Bell did on that very same Valentine's Day.

Once Bell returned from Washington, in what seemed like a moment of divine inspiration, he sketched in his notebook the same device that had appeared in Gray's application. In his investigation, Seth Shulman overlaid the image from Gray's patent application with Bell's sketch. They are almost identical (Shulman 2008).

Theft?

The trouble all started when Bell fell in love with a vivacious teenager called Mabel Gardiner Hubbard. In his attempts to woo her, Bell convinced her rich father, Gardiner Hubbard, an attorney specialising in patents, to invest in a scheme to create a device to send sounds over telegraph lines (Shulman 2008). This device would be the world's first harmonic telegraph.

Hubbard agreed to the investment. However, until he saw progress, he would not permit a marriage. Bell's biographer, Charlotte Gray, said that Hubbard's threat was clear. If Bell didn't buck up his ideas, Hubbard would end their partnership and slam the door of his house, and any chance of marrying Mabel, straight in his face (Gray 2006). What went wrong for Bell? The problem was that Bell's practical skills didn't match his conceptual brilliance. After a year's worth of work, he had nothing to show for his efforts or for Hubbard's money.

As for Hubbard, Cheung and Brach are unequivocal that he was ruthless and a cheat (2020). According to them, in February 1876, Hubbard found out that Elisha Gray had submitted an application to the Patent Office for 'an apparatus using telegraphic means to transmit and receive sounds'. Hubbard then took it upon himself to gather Bell's notes and submit a half-baked patent application on Bell's behalf. Once filed, Hubbard bribed a Patent Office employee to record that Bell's patent had arrived earlier in the day than Gray's.

Ten days later, Hubbard arranged a secret meeting between Bell and another corrupt Patent Office employee, this time an examiner. The examiner let Bell see Gray's idea of converting sounds to electrical signals. Bell then altered his already filed application with the key idea that would make the harmonic telegraph work. He copied the diagram into his notes on 9 March, which, Shulman says, explains the mystery of their similarity (2008).

The device

In Elisha Gray's picture, a man's face speaks into what looks like a little drum. The little drum is a diaphragm. A conducting pin hangs off it. The other end of the pin dangles in electrically conductive fluid. Speaking into the diaphragm caused unique vibrations and therefore tiny, and equally unique, movements of the pin. This movement altered the resistance of the current in the fluid and thus created a unique electrical signal.

On the other end of the line, an electromagnet and another diaphragm made up what was called a 'speaker'. The electromagnet converted the electrical signal back into the original vibrations that had gone in. In this way the harmonic telegraph transmitted the other person's voice, a dog barking or – if Shulman is correct – the sound of Bell's shameful silence. In this way, something in the air, assisted by the circuitry, transformed into an electrical current that was later transformed back into sound.

Enter Edison

Gray's design was elegant. It was also impractical. It took a young Thomas Edison and a masterstroke to make it both practical and beautiful.

After Bell's patent application, Western Union built their own telephone system. They assumed Bell and Hubbard wouldn't dare challenge them in court. Besides, with Gray on their staff, Western Union were not convinced that Bell did all the work to invent the telephone. In other words, they thought they could get away with infringing on Bell's patent. Given what was at stake, it was a calculated risk worth taking.

Western Union hired Edison, who was still only in his twenties, to come up with a more practical design to replace the pin dangling in the solution. His device came with tightly packed carbon elements in what he called a 'carbon button'. When soundwaves entered the button, they pushed the carbon elements closer together. The change in the carbon elements altered their resistance, just as the moving pin had altered the resistance of the electrified fluid. However, unlike the pin in fluid, Edison's carbon buttons were small and resilient. They easily fitted, and could easily be fitted into, devices that were much smaller than the telephone equipment of the time. The carbon button was the key component that allowed Edison to create for Western Union a small or 'micro telephone'. The name microphone stuck.

A key difference

Many systems that use electricity are similar in that something happens on one end of the line that's reversed on the other. A generator converts motion into electricity, which is then sent down the wire, where it's changed back into motion by a motor (or where it's converted into heat, light, cold or sounds on the radio). This is also true for the telegraph and the telephone. In both systems there's a sender, a wire down which messages are transmitted at speeds approaching that of light, and a receiver.

However, within the shadow of the telegraph and telephone's similarity lies a key difference. Human operators encoded and decoded messages at each end of the telegraph line. The sending operator encoded the message according to a predefined scheme, Morse's code. The receiving operator then deciphered it according to the same scheme. In this way the operators were human computers.

The telephone, however, had no operators at all. The person sending the message spoke and, due to the electrical machinery on the other end, which reverses the process of converting sound into electricity, the receiver heard the message directly and exactly as it was spoken.

The signal used in the telegraph was digital. A digital signal carries data that represents something else. In the case of the telegraph, the data was dots and dashes and they represented letters.

The most common form of digital data is binary data. Binary data takes only one of two forms, 0 or 1, as per the binary number system. As the telegraph's dots and dashes demonstrated, binary data is well suited to electrical means of communications because a 0 can be represented by a short burst of electricity and a 1 by a longer burst. In digital computers, 0 is represented by the absence of a current and a 1 by its presence. A single binary digit is called a bit – **b**inary dig**it**. A lot of binary numbers together, such as the whole of a message sent with the telegraph, is called a bitstream.

The telephone system was not digital. The phone lines did not carry data that represented another form of information so there was no need for an encoding scheme. Instead, the electrical signal was as complex as the soundwaves from which it was produced. All that was needed on the receiving side was equipment that could turn the electrical signal back into the sounds that had gone in. The electrical signal used in the telephone system was therefore analogous to the sounds it represented. This is why such signals are called analogue signals.

Digital signals degrade more slowly than analogue signals and can therefore travel longer distances. In addition to that, because with binary representation a current is either present or not, digital signals are easier to amplify. Analogue signals, on the other hand, degrade more rapidly and can only be amplified a few times before degenerating into garbage. This is why the telephone did not instantly take over the telegraph. It only worked locally. It wasn't until 1915, after a decent amplifier was invented, that the first telephone call was made between New York and San Francisco.

The most important concept in computing just crossed our paths. Electricity can be converted to sound, heat, light and motion. But it can also be converted into bits in a digital stream. Electricity can, in other words, be used to represent information.

Where to now?

In 1879, Thomas Edison made a breakthrough that let him 'divide' light across a grid of electricity. This is one of the most important moments in the history of the world because it brought forth the electricity grid. However, it was also the first time the Edison effect was observed. The Edison effect is what makes computers as we know them work.

Away from his workbenches, the culture at Edison's Menlo Park facility encouraged creativity, used a technique that modern programmers call evolutionary prototyping, had no bureaucracy and not much of a hierarchy, either. Like the team at Amazon in the 1990s, the team at Menlo Park took failure seriously.

Maslow would have described Edison's system of management as 'humanistic'. It is to that system, and the invention factory it came from, that we now turn.

Chapter 3

The invention factory

By 1869, the great inventions of the First Industrial Revolution – the steamship, the train and the telegraph – really had annihilated the barriers between space and time. The job of the Second Industrial Revolution was to pick up where the first had left off.

As the 1880s began, this was about the last mile. Trains couldn't get passengers all the way to their front doors. That changed when Henry Ford's Model T arrived. The telegraph had shown that electricity could be transported and at the point of consumption, with the click-clacking of the sounder, converted into information. It fell to the telephone, however, to get messages into houses and solve the last mile problem for communications.

Along with the internal combustion engine, wireless communication and the telephone, light bulbs and the system to get electricity to power them were defining inventions of the Second Industrial Revolution. This is, of course, where Thomas Edison comes back into the story.

Two problems

Edison needed to solve two problems. First of all, the incandescent light bulb needed perfecting. It had been around for years but was rubbish. Second, once perfected, or at least on the way to perfection, Edison had to create a vast system of illumination, the components of which needed inventing, manufacturing and, with the help of brown envelopes stuffed with cash to bribe city officials, installing (Morris 2019).

One insight

In 1879, Edison took to looking through a microscope at different materials such as platinum and iridium, as they incandesced inside light bulbs. This blinding work, which brought to his eyes 'the pains of hell', showed that just before it burned out, the filament popped (Morris 2019). The popping happened just before the oxygen leaking out of the heated metals ignited and burned through the filament, thus bringing the vacuum to an end.

At the time of these blinding observations, Edison realised that if the wire in the filament, a word he repurposed from botany, was wound tighter, then it gave off more light. But more wire also meant more resistance in the circuit. The filament, as a resistor, therefore *increased* the dissipation of light while the thickness of the conducting wires 'feeding' them *decreased* (Morris 2019). This insight made the bulb and the grid possible.

How? Electricity flows like water through a pipe. A resistor slows down the flow. Low-resistance bulbs needed thick pipes to bring enough 'water' into them. This is why Edison likened unresisted electricity to water rushing through pipes and emptying the reservoir upstate (Morris 2019). This explains why low-resistance bulbs required a prohibitively expensive amount of conducting cables and why, as late as the 1880s, a grid of electricity was impossible.

High-resistance bulbs, on the other hand, requiring much less 'water', meant the impossible became possible: a grid that delivered electricity. This happened in a flash of inspiration. It was the spark of genius that made the grid possible, and Edison was alone in having it.

Collective amnesia

Shortly after these blinding observations, in October 1879, and despite his obsession with it, Edison conceded that platinum as a filament was a dead end. He returned to carbonised materials such as bits of fishing line or coconut twine. To the delight of all involved, one of the filaments glowed for 13 hours.

The exact moment these ideas came together is not clear. Brad Stone (2021) said that nobody remembered what happened in the early days at Amazon, either, such was the febrile pace of development. Those involved, he said, suffered a form of collective amnesia.

The same collective amnesia happened at Menlo Park. Despite Edison's prodigious note taking, the filament's exact moment of discovery vanished into history and, like those heady days at Amazon just over 100 years later, dissolved into hearsay before hardening into myth.

However, what we do know is that on the evening of Tuesday 21 October and the wee hours of Wednesday 22nd, the filament kept glowing. On that fateful Wednesday morning, for all the doubts of restless financiers, for all the mocking from Victorian England of this audacious and beardless huckster, Edison and his team found a filament that took the heat and rendered beautiful light (Schewe 2007). This was the moment that Edison's team knew the 'old man' was going to do it. He was 33.

Creativity meets capital

It was thus, cooped up at Menlo Park with throbbing eyes, that Edison and his team perfected the high-resistance light bulb. This freed the team to accelerate their work into new and improved steam-powered generators, transmission cables, circuit breakers, conduits, switches and the yet-to-be-invented electricity meter, which measured how much future customers consumed and would therefore pay.

On top of all that, to avoid tangling with the telephone and telegraph lines that had sprouted around New York like weeds, Edison had to, and wanted to, bury his cables underground. To pull that off, he needed rodent-proof cables and waterproof seals, both of which needed inventing, and those envelopes stuffed with cash for the city officials.

All of this pointed to an inescapable fact: Edison's system required a lot of start-up capital. Where would it come from? Because of their recent experience in court with the Bell Telephone Company, where they argued about that contraption with the pin hanging in fluid, Edison's backers, John Pierpont Morgan and Western Union, vacillated.

The demonstration

To assuage their doubts, on New Year's Eve, 1879, just weeks after he cracked the filament problem and Western Union settled with Bell, Edison invited his backers (and the press) to Menlo Park. They arrived at night. Edison stood in the dark. In the winter's air, his breath must have been visible against his silhouette. With the flip of a switch, the grounds, which were powdered with fresh snow, lit up. The audience gasped. Edison threw another switch. A series of blinking bulbs sprang to life. The astonished guests burst into laughter. They had witnessed the turning on of the world's first Christmas lights.

The miracle

The demonstration was the beginning of the hard work. A few frantic years of creativity followed. During that period, Thomas Edison combined the invention of a practical light bulb that could be manufactured at scale with the invention of electricity generation and distribution to houses, factories and street lights (Gordon 2016).

Kerosene lamps no longer had to be cared for or 'fed'. Light had nothing more to do with dripping oil, trimming wicks or the consumption of oxygen. Respiratory illnesses would no longer be aggravated by fumes trapped indoors. Pollution shifted from houses and factories to the point of electricity's generation, at the power plant. The user and the source of electricity were divorced in space and time. Those who had electricity generators in their gardens, like J P Morgan, no longer suffered from burned carpets and singed furniture (Schewe 2007).

Applications

The miracle, though, was not limited to light. A clue appeared in Edison's patent application number 369,280. It said that his invention was 'a system for the generation, supply and consumption of either light, or power'. Edison intuited that houses wired up for light would be used for other yet-to-be-invented applications. This was soon shown to be true.

By the time the 1930s rolled around, electricity was converted into entertainment in the form of sound by the almost ubiquitous radio. It was converted into motion by electric trains, into cold by refrigerators and later into heartbeats by pacemakers.

The ability to plug applications into the grid was remarkable, so we can therefore add flexibility to cleanliness and silence (Smil 2005). A similar flexibility reemerged in the 1990s with the internet and again in the 2010s with the cloud.

The unforeseen applications of the grid meant that Edison learned the same lesson as Cooke and Wheatstone: the destiny of a general-purpose technology lies not in the hands of its creator but its users, who seek out applications that were hitherto unimaginable.

The man with many faces

What about Edison the man? Who was he and how did he redirect the trajectory of the human race? His character is worth unpacking because it teaches us so much about how to succeed with systems development and innovation right now, in 2025.

Edison the creator

Edison had three skills, which are rarely found in the same person. First of all, he was a great systems integrator. Many of the technologies the grid depended on were already available, albeit in a poor state of development. What was missing was the ability to bring them together into a coherent whole. Edison did that, envisioning the whole grid at the same time as he studied its smallest component, the filament in the bulb.

Second, Edison was a great inventor. Lots of components for the system he envisioned had not yet been invented. What began in his mind would reappear in the real world just as the carbon button, the Quadruplex telegraph and the phonograph once had.

Finally, Edison brought all the components of the grid, some of which had not yet been invented, into a coherent whole and *then went on to improve almost every aspect of the individual parts*. Edison was, in other words, a tinkerer extraordinaire. The ability to see the whole, and in doing so connect all the dots, invent and incrementally improve components are vital to innovation.

Edison the manager

Menlo Park, in New Jersey, was an unfinished development of about 40 houses located at the junction of the Philadelphia Turnpike and the Pennsylvania Railroad (Morris 2019). There were a few houses and, a short distance away, Edison's workshop. Soon after Edison arrived, Menlo Park became known as 'the invention factory'.

Two teams lived at Menlo Park. The first team experimented. The second manufactured and prototyped the equipment that the experimenters needed. Organised communally and democratically, Menlo Park depended not on hierarchy or paternal management but instead on a strong vision and Edison in a role similar to that of an orchestra's conductor.

It was in this way that Thomas Edison got the best out of his technologists, who were using the technology of the day to invent the technologies of tomorrow. The art of getting the best out of people is called management. Edison was good at it.

Edison the innovator

Technologies changed in the last century. The process of innovation did not. Innovation is about systematically seeking solutions to problems, which themselves may not be known. This leads to the peculiar situation where a solution, such as the battery, is known before the problem it was destined to solve appears. Innovation, therefore, explores the problem and solution space at the same time. This process of discovery progresses through experimentation, which is a posh way of saying that the process of discovery progresses through trial and error.

Thomas Edison's method of trial and error became known as the Edison approach. It was useful when there was no prior art or theory. For example, when it came to finding the best filament for the incandescent light, there was no general knowledge about which material incandesced best, so a systematic search was the only logical approach.

Edison's approach was scientific but it wasn't science. He didn't design experiments for other scientists to repeat in their labs. Instead he generated new knowledge for whatever problem he was solving at a given moment. One of the ways he generated new knowledge was by increasing the complexity of the environments in which his inventions operated. For example, he wired up Menlo Park, which he demonstrated to his visitors on that fateful New Year's Eve. The lessons he learned when wiring up Menlo Park were then fed into his project to wire up New York.

The evolution of a system based on feedback from the environment is something computer programmers call *evolutionary prototyping*. Evolutionary prototyping involves developing a system, deploying it to a lifelike environment, like Edison did at Menlo Park, and then studying how the system behaves or fails to behave. Evolutionary prototyping allowed Edison to migrate ideas from his imagination and into the real world. In the process, he learned more about both.

Edison the failure

Edison knew that a successful innovation was the sum of its failures (Wills 2019). The more failures that occur in the development of an innovation, the more successful it will be when it makes contact with the real world and its unpredictable users. Conversely, innovations that do not fail during their development are doomed to fail, sometimes catastrophically, when they eventually do collide with the real world. Edison understood this and that explains why he took failure seriously.

Lessons from Edison

Edison taught us all a few things about systems and their development. Before we leave this chapter and the 19th century behind us, let's take a look at them.

- **The creators of a system do not dictate its destiny.** A recurring theme in the history of technology, one that returned in the 1990s with the internet and again in the 2010s with the cloud, is that it's not the creator of a technology who dictates its destiny. It is its users. The range of applications for the grid, many of which Edison could not envision, shaped what happened after the grid was invented.
- **One change can change everything.** Edison demonstrated the nature of systems and explained why they're so notoriously hard to debug. A change in one component can change the behaviour of the entire system. The changes made to the light bulb in turn changed the thickness of the conducting wires. That one insight made the grid possible.
- **The sum total of failures leads to success.** Edison took failure seriously because he knew that a successful innovation was the sum total of its failures in development. Because a change in one component had the potential to change the entire system, he also knew that a process of trial-and-error tinkering is essential to systems development.

- **Failure is the norm in innovation, so failure must be normalised.** Edison did this with a system of management that created space to discuss failure, to brainstorm and to share results. It goes wrong for managers, for example those using cloud technologies, when failure isn't normalised. When that happens, their instinct to avoid risks overwhelms their instinct to innovate.

We will return to these features of systems and their development throughout this book.

END OF PART 1

Not wanting to be outdone by his uncle and his frogs' legs, Luigi Galvani's nephew, Giovanni Aldini, visited London in 1803. He conducted an experiment on the body of a recently hanged criminal. Observers gasped in horror as 'an eye opened, a fist clenched, a leg twitched' (Winterson 2022).

A few decades later, as the Enlightenment was drawing to a close and the First Industrial Revolution was starting to pick up pace, almost certainly with Aldini's experiments in mind, Mary Shelley wrote: 'By the glimmer of the half-extinguished light, I saw the dull yellow eye of the creature open; it breathed hard, and a convulsive motion agitated its limbs.'

This was the moment that the creature in her book, *Frankenstein, or, The Modern Prometheus*, came to life or, maybe more accurately, came back to life (Shelley 2012).

Just 79 years later, in 1897, Bram Stoker published *Dracula*. In Stoker's yarn, Dracula uses steamships, Edison's light bulbs, Cooke and Wheatstone's telegraph and the railways. The intuitiveness of these technologies made the modern world easy for Dracula to navigate. Their intuitiveness explains their rapid adoption – what was good for an old vampire, it turned out, was good for everybody else.

The 19th century started with *Frankenstein* and ended with *Dracula*. They each describe a radically different world, despite there being only 79 years between their publications. As Jeanette Winterson wrote in her 2022 book, *12 Bytes*, 'Each sits like a bookend at either end of the century' that saw 'more change than any other period of history'. It was also during this period, especially towards the end with Edison's work at Menlo Park, when a number of themes important to the history of computing emerged. Before we leave Part 1, let's take a quick look at them.

Themes

The first theme is that new technologies are often solutions looking for a problem. That was true for the battery and the telegraph. Applications of a technology emerge from their use in the real world. Cooke and Wheatstone did not predict their telegraph would be applied to crime busting. Phillip Schewe compares this to Kierkegaard's observation that life can only be understood backwards but it must be lived forward; humans build machines and then work out what to do with them (2007).

A second theme is that, until they see it working, people struggle to understand a technology. Vivid demonstrations can cut through the noise of the day. These demonstrations are helpful during the crossover decades. The crossover decades are what happens when a new technology comes online but people have yet to update their frames of reference.

It's easy to poke fun at our ancestors who lived in earlier crossover decades. After all, they envisioned no practical use for the telegraph. Not long after that, Western Union said that no sane person would ever want to talk into a telephone. One hundred years later, in the 1990s, booksellers laughed at Amazon. Who in their right minds, they wondered, would buy books using a computer?

We should not make fun of our ancestors. The age of artificial intelligence is upon us right now. We are ourselves in the crossover decades and, since the effects of artificial intelligence will be more transformative than anything that came before it, we should be panicking. Instead we seem as daft as those politicians that poor Samuel Morse had to put up with. What will our children laugh at us about?

A third theme was systems. We learned that a change in one component, like the change Edison made when he wound the filament tight and increased its resistance, can alter the whole system. It's for this reason that Edison took failure seriously. He knew that a successful innovation was the sum of its failures. The world is full of businesses run by managers who want to use the cloud. However, they haven't learned that success will come from day-to-day failures. If their instinct to reduce risks overwhelms their instinct to innovate, they will drive out the source of their eventual success. Edison's system of management, with its democratic processes and culture of honesty, was designed to stop that happening. Menlo Park therefore foreshadowed the great research organisations of the last century and Edison's system looked a lot like what would later be called humanistic management. Edison's work at Menlo Park imparts valuable lessons for those who are about to start playing in what, to them, is the brave new world of cloud computing.

The final theme was computers. We learned a few things about them, too. By comparing the telegraph and telephone, for example, we understood the difference between analogue and digital signals. We also learned from the telegraph the most important concept in computing: electricity can be used to represent information. This in turn taught us that information can be communicated at the speed of light. Since that became true, people at the turn of the last century soon started to wonder, maybe, just maybe, information was also manipulable at the speed of light, too.

Where to now?

What started on Edison's workbench was about to bring about the end of the age of industry and usher in a new age of information. The Second Industrial Revolution was, in other words, about to give birth to the Third. It is to this new age of information and the speed of light manipulation of electricity that we now turn.

PART 2

ELECTRONS AND NUCLEAR BOMBS

Chapter 4

The peculiar birth of a peculiar company

Years before the electrification of society, before the telephone arrived, but a few years after Morse invented the telegraph and right in between the publication of *Frankenstein* and *Dracula*, a French novelist, writing with remarkable foresight, saw something on the horizon of humanity's future. And it wasn't a telegraph pole.

The two principal characters in Jules Verne's *Twenty Thousand Leagues Under the Sea* (1870) are Captain Nemo and Professor Pierre Aronnax. On a tour of Nemo's submarine, the *Nautilus*, the captain tells Aronnax: 'There's a powerful, obedient, swift and effortless force that can be bent to any use and which reigns supreme aboard my vessel. It does everything. It lights me, it warms me, it's the soul of my mechanical equipment. This force is electricity.'

The bending of electricity to Nemo's whim comes from a combination of electrical and electronic engineering. Verne, therefore, not only predicted the capturing of Zeus's thunderbolts and their subsequent bottling in a grid of restless energy; he also predicted that unknowable components that could manipulate electricity would one day be invented and that those components would be small enough to fit inside vehicles that they would control.

Edison's effect

Verne was accurate with this prediction. Not long after the electrification of society began, a sub-discipline of electrical engineering called electronic engineering was born. Whereas electrical engineering was about Zeus's thunderbolts and the machinery in which they were encased, such as wires, switches, capacitors and inductors, electronic engineering was about manipulating electricity in order to carry out tasks so advanced they would seem magical.

The story of this magical manipulation started life on Edison's workbench. He noticed, when he was half blinding himself with those light bulbs, that after a week or two of incandescing, black soot would gather on the inside of the glass. This was impossible to clean and, if the soot got thick enough, the light would not escape from the bulb. This was a mystery because soot comes from fire and fires don't burn without oxygen, and there was none of that near the vacuum-sealed filament.

Things became more mysterious when a blue, fluorescent light that responded to magnetism appeared near the clamps. Imagine that, bending light with a magnet. Edison intuited that the soot and the blue light were related but he didn't have time to do any serious investigation. He registered a patent (of course) and went back to bringing the electricity grid to life. His observation was filed away in history as Edison's effect. It remained there for nearly 20 years until a man called J J Thomson came along.

The birth of electronics

Through a handful of clever experiments, Thomson was able to show that no matter what the filament was made of, something unrelated to its physical properties was flying around inside those light bulbs. That something, he correctly postulated, was a thousand times smaller than an atom. He had discovered the most basic carrier of electrical current. It was later given a name: the electron.

A few years after Thomson's experiments and more than 20

years since Edison noticed it, in 1904 an Englishman called John Ambrose Fleming, who had worked at Edison's London branch and was familiar with his effect, invented something called the Fleming valve.

The Fleming valve was a light bulb with a copper plate inside it, exactly like the one Edison had put inside his bulbs to try to shield the glass from the soot. If that plate was charged one way, electrons flowed across the vacuum. If it was charged another way, they wouldn't flow. The 'stream' of electrons could be switched on and off. The Fleming valve was therefore a tap for electrons. Because it had two things inside it – the plate and the filament – in the United States the Fleming valve became known as the diode.

What had started at Menlo Park was finished by J J Thomson's experimental observations and exemplified in Fleming's valve. The age of electronics had arrived.

The triode

In 1906, Fleming's souped-up light bulb got another modification (Reid 1985). An extra bit of wire, called an electrode, was inserted between the little copper plate and the filament. This was a brilliant addition. Instead of switching the flow of electrons on and off like the Fleming valve did, because of this extra bit of wire, this new device could control the size of the flow in the same way that a dimmer switch controls light or a dial controls the volume of a speaker. The flow of electrons could be slow, fast or anything in between, including fully on or fully off.

Because this new version of the souped-up light bulb had a third component, that extra bit of wire, it was called a triode. Later, its erratic behaviour, due to gases in the tube, would be corrected by advances in instruments that allowed researchers to create the 'hard valve' or the 'vacuum tube' (Landes 2003).

Signal amplification

The vacuum tube, it turned out, was a bit of bottled magic. Described as 'the truest "little giant" in all of history' (Landes 2003), when hooked up to an antenna the tube amplified a radio signal enough to drive a loudspeaker (Reid 1985). Its amplification abilities destined it to be crucial to the development of television, radar and telecommunications. It was therefore of great interest to the American Telephone and Telegraph Company (AT&T).

One system, universal service

After Alexander Graham Bell's win in court against Western Union, things went downhill for the Bell Telephone Company. In 1894, Bell's patent expired, which led to an explosion of competitors. Then, in 1899, to get around a law in New York that restricted its growth, the Bell Telephone Company was acquired by one of its own subsidiaries, and the whole operation became AT&T.

The legal jostling got them out of New York's laws but did nothing for their reputation. AT&T were a public menace (Gertner 2012). When they could not defeat their competitors in court, they would resort to sabotaging their lines. Behind the scenes, AT&T strengthened their position by buying equipment suppliers and then made sure they wouldn't supply their competitors, whom they also refused to work with through a policy of noncompliance (Gertner 2012). This is why some users found themselves in the maddening situation of having three telephones in one building, one for each service provider.

At the same time, AT&T's customers received awful service and those in the countryside, always the very last to benefit from advancements of the Industrial Revolutions, were stuck on a party line, if they had any service at all (Gordon 2016). In short, the company's practices were aggressive and unethical and, from its very inception, after the patent battle with Western Union and Elisha Gray, litigious (Gertner 2012).

AT&T's downward spiral was eventually arrested by Ted Vail's 1907 return. Vail was previously the general manager who defended Bell's patents against Western Union. For his troubles, he was ousted after the court case (Gertner 2012).

Vail came to see that AT&T would work better, be better for everyday citizens and of course be more profitable if it was allowed to be a natural monopoly. AT&T argued its case with the government. The government agreed, but there were conditions. One of them was that AT&T had to work with the government to set its prices and profitability. Once agreed, Vail got busy bringing to life his vision of using cutting-edge technology to deliver one system with universal service *that actually worked*.

The vacuum tube gets plugged in

At the time, the system did not work. Vail did two things to change that. First of all, he would acquire, rather than fight, AT&T's competitors. Second, he hunted for technology that would give AT&T an edge. This is how he found the triode and its patent holder, Lee de Forest. In a remarkable case of the inventor not knowing what they had invented, de Forest agreed to license it to AT&T for the bargain basement price of $50,000. AT&T then improved its performance and helped change it into the hard valve or vacuum tube.

In 1915, AT&T placed the vacuum tube at strategic positions along a coast-to-coast line. The ageing Alexander Graham Bell was in New York. His former assistant, Thomas Watson, was in San Francisco. They repeated their 1886 experiment when, using their pin in fluid, they had made the world's first phone call. That day, Bell said, 'Mr Watson, come here, I want to see you.' This time round, Watson replied that it would take him a week to get there.

Bell Labs

After the breakthrough that came from the vacuum tube, Ted Vail's vision of a singular system built from the best technologies started to look as if it could be made real. Not long after that, he sailed off into retirement, but AT&T kept his vision alive when, in 1924, it managed to kill two birds with one stone.

AT&T needed a way to systematically develop new technology that would bring the telephone to everybody, thus making it a genuinely universal service. That was the first bird.

AT&T also had to make sure that profits were kept low. That was hard to do since they had achieved their goal and become a natural monopoly. But if they were to avoid the wrath of the monopoly-busting government, they needed to eradicate their profits. That was the second bird.

The stone that killed them both was a company that AT&T founded called Bell Labs. Bell Labs was fully dependent on AT&T. Its stock was not for sale (Gertner 2012). It would never be profitable. Its managers and those at AT&T weighed every decision against what they thought the Department of Justice might do. Bell Labs was, in other words, a weird company.

Despite its weirdness, though, one thing was clear: Bell Labs' remit was broad and its purpose singular. It would look into a massive range of ideas, from woodpecker-proof telephone poles to automated switching systems, but it would do so with one goal in mind: to improve the telephone service while driving down the cost of a system that was vital to the standard of living of millions of Americans. This was how AT&T burned its profits for the good of humankind.

Where to next

As well as amplification, vacuum tubes had a secondary and much easier to understand function: they were a fast switch. When it was off, the vacuum tube could represent the binary number 0; when it was on, the binary number 1. As well as being used to amplify analogue signals, as they had done between New York and San Francisco, vacuum tubes were therefore destined to be put to use inside the world's first electronic computers. Lee de Forest had no idea the chain of events that he'd set in motion when he stuck that extra wire inside Fleming's valve.

Chapter 5

The computer

Before mechanical counting machines were invented, professional mathematicians and scientists hired boys or young men to work as computers. The job of a computer was to perform long and tedious computations. Only in exceptional cases did women work as computers. Sometimes working as a computer was a stepping stone to greater things. For example, Johannes Kepler started his scientific career as a computer, assisting more experienced scientists. Often separate teams of computers worked on the same calculations so their results could be compared with each other for correctness.

What on earth were these calculations that required so many human computers? They worked out the trajectory of objects in space or the speed at which a train could travel before its passengers' heads exploded. Inside Victorian firms, they were the mind-numbing and repetitive computations that hundreds of clerks performed day in and day out.

Computing (and clerking) was laborious, expensive, tedious, error prone, repetitive, badly paid and, to its own detriment, dominated by men. The telegraph had facilitated an escape from the tyranny of proximity but nothing freed humanity from the tyranny of all these numbers and their mind-numbing crunching. Thankfully, one day that started to change.

The promise of electronic computers

Computers and the scientists they worked for were assisted by the invention of rudimentary mechanical counting machines. This helped them to partially escape the drudgery of their computational work. Further escape was promised with the arrival of the vacuum tube.

Like all bulbs, the vacuum tube could be easily switched on and off. If you had enough of them, because they can be used to represent information, they could be arranged into a circuit that carried out basic computations. Unlike their mechanical or human versions, electronic computers were fast. Unlike a vacuum tube, a mechanical switch cannot be flipped on and off 10,000 times a second (Reid 1985). It was this fast switching that promised an escape from the awful drudgery of manual computations.

ENIAC

The broad and creative work into electronic computers, both in the United States and Europe, between the invention of the vacuum tube and the end of the Second World War in 1945, culminated with machines such as the University of Pennsylvania's electronic numerical integrator and computer (ENIAC).

One of the world's fastest computers, without a single moving part, ENIAC performed its calculations at the speed of the electron and not the speed of mechanical wheels and ratchets.

However, ENIAC and computers like it were somewhat anomalous in the late 1940s. Digital technologies such as the telegraph and Morse's code were old hat. Using electronics to create digital behaviour was at that point the exception and not the norm (Dyson 2012). Television, the telephone, radio and vinyl records were all analogue technologies. It was these devices and the research into them that sat at the cutting edge of electronics. Why did the ENIAC's designers break the current thinking around electronics?

The vacuum tube was about controlling the flow of electrons to amplify a signal or blast them onto a screen, as in radar systems, where their charge temporarily lit up as information the operator understood. Used in this way, the control of electrons was remarkably powerful but could not help at all with automated computations. This was a problem because, in 1944, the military had a very serious computational problem to solve. For that, they would need to use electrons to represent information, exactly as they were once used to represent letters with Morse's dots and dashes. Digital electronic technologies were about to join the party their analogue cousins were already at.

ENIAC's raison d'être

Out in the theatre of battle, firing tables contained the trajectories along which anti-aircraft guns should fire their shells. Back in the United States, teams at the US Army's proving ground in Aberdeen, Maryland, calculated the tables. A couple of rudimentary (but not electronic) computing machines and extra human computers working over at the University of Pennsylvania helped with the effort. A single human computer could calculate *one* trajectory in about 12 hours. Working with mechanical computers, it could be done in about 20 minutes (Dyson 2012).

This meant that, working in double shifts, the teams at the proving ground and the university could just about produce a single firing table in a month. Unfortunately, shells and planes continually improved and each improvement required a new firing table. In 1944, those improvements led to the team at the proving ground receiving requests for six new firing tables per day. The humans couldn't keep up (Dyson 2012).

ENIAC was designed from the ground up to solve this computational problem. Electrons would still flow like water but, inside computing machines, they would mean something. The charges they formed represented data and the instructions that would transform that data into information with meaning. Nowadays, this sort of transformation is about changing strokes on the keyboard into orders for books or heartbeat data collected on smartwatches into dietary and exercise advice. Back then, for the ENIAC and its human operators, the transformation changed the flight data of Allied shells and German planes into firing tables. They in turn gave the Allied gunners a chance to shoot the latter down with the former.

The ENIAC team succeeded in its mission, but not before the war reached its end. The ENIAC therefore never did compute in anger but it nevertheless did in 15 minutes what a whole team of human computers used to take a week to do. This is why a 1946 *New York Times* article said the ENIAC applied electronic speeds 'to mathematical tasks hitherto too difficult and cumbersome for solution' (Kennedy Jr 1954).

A big problem

The awesome potential of electronic computers had been proven, albeit within a narrow range of applications, during the Second World War. ENIAC, soon after, showed what theoretically could be done with computing machines. Then, in 1951, ENIAC's successor, UNIVAC (Universal Automatic Computer), demonstrated the computer's further usefulness when it predicted Dwight Eisenhower's landslide victory in that year's presidential election. The UNIVAC, seemingly, liked Ike, too.

However, for all the improvements in the postwar years, the escape from the tyranny of numbers that vacuum tubes promised never fully materialised. Electronic computers, like their mechanical predecessors, shared the same serious drawback. They were massive. People joked that the ENIAC consumed so much electricity that when it was switched on, the lights in Philadelphia dimmed (Gordon 2016). The B-52 bomber, which came into operation in 1952, is bigger than the Airbus A380 and even it could not carry the ENIAC.

Nowadays, a powerful computer fits in a pocket or backpack but back then a person – in fact, hundreds of people – could fit inside a computer. Unless those machines could be miniaturised, they would never fit on a plane or submarine and could never be installed in offices where they could relieve the mind-numbing number crunching that went on inside them. In 1945, Jules Verne's vision was tantalisingly within sight but still miles and miles out of reach. Unless something could be done about their size, practical electronic computers were doomed to remain a figment of humankind's imagination.

The transistor

It was at this exact moment that Bell Labs and their mind-bogglingly wide range of research came back into the picture. By 1947, AT&T's public relations team had not only done a good job of rewriting their own history but its managers had ensured that Bell Labs was properly organised and crammed full of some of the greatest minds in America. The miniaturisation of electronic computers, and therefore the promise of cat memes and online bookstores, was about to be thrown a lifeline by the telephone company.

More accurately, miniaturisation was about to be thrown a lifeline by Walter Brattain and John Bardeen. In 1947, Brattain and Bardeen invented the transistor, the world's first electronic device made from a semiconducting material. The transistor could both amplify an electrical current and act as a switch. In other words, transistors could do exactly what vacuum tubes did. A vacuum tube, however, could never be made much smaller than any other light bulb, which meant that a decent computer could never be made much smaller than a house. Transistors had no such limitation.

The fabulous midget

However, this was not the only thing that made the transistor fabulous. Other than their size, vacuum tubes had three other serious drawbacks, all of which stemmed from the inescapable fact that they were just light bulbs.

First of all, because they got hot, if vacuum tubes were packed too tightly or switched on for too long, they would melt the machinery they were part of. This is why, a few years later, the IBM 704, housed in Building 26 at MIT, had three engineers assigned to it who sprang into action, like Pavlov's proverbial dog, when the air conditioning broke. Their job was to undo the panels quickly before the 704's innards melted (Levy 2010).

Second, the light that vacuum tubes emitted, including the strange fluorescent glow that Edison had seen and which emanated from the back of my grandma's radio, attracted moths. They got into the machinery and short-circuited it. This is why computer operators were always *debugging* their machines. Grace Hopper, an early pioneer of software development and later one of the US Navy's few female admirals, pinned a dead moth in her lab book and noted 'Relay #70 Panel F (moth) in relay'.

Finally, all filaments eventually burn out the bulb because the tremendous outside force on the glass that protects the vacuum on the inside cracks. The ENIAC used high-quality tubes but there were so many of them – 18,000 – that one would blow every few seconds. Platoons of soldiers were deployed at strategic positions with grocery bags full of vacuum tubes so they could replace them as they popped (Reid 1985).

The transistor brought an end to this madness. The 'fabulous midget', as one academic called it, was a like-for-like replacement for the vacuum tube, except it was small and had low power needs. This is why the big and power-hungry radio was quickly replaced by a battery-powered transistor radio that could fit into a shirt pocket. Once the fabulous midget had arrived, the prize that Jules Verne sketched out was not only in sight but was well and truly within reach. The race to miniaturise computing machines was on.

The tyranny of numbers, again

That race got off to a slow start. It took an agonising 12 years to reach its next milestone. In those 12 years, something had gone very wrong for the fabulous transistor. The computer industry had rapidly embraced it, and in the 1950s designers were having stunning visions of a computing machine that could be made as small as a car. At the same time, the military needed low power and resilient little guidance computers for their ballistic missiles, which they thought could be built from transistors. What the transistor did for the radio surely could be done for computers and control devices like the ones on the Nautilus? Actually, no. Amplifiers are simple circuits. They use a vacuum tube or transistor, have a power source, a resistor and a capacitor. That's about it. Pull the back off a pocket radio from 1954 or take a peek on Wikipedia, and you'll see a handful of transistors, resistors and capacitors all wired up *by hand*.

Computing machines or control devices needed tens of thousands of components. The connections between the components, however, could easily be ten times their number. All those connections, as they were in the transistor radio, needed soldering *by hand*.

What could be done for the radio – the soldering together of components made from a semiconductor – could not be done for complex machines like computers. The electronics industry called this the 'interconnectedness problem' or the 'tyranny of numbers'. This was a serious problem and it meant that, beyond the transistor radio and hearing aids, there were hardly any practical applications for the transistor.

The monolithic idea

When all hope must have seemed lost, in July 1958 Jack Kilby hit on the solution to the interconnectedness problem. Not long after that, Robert Noyce independently arrived at the same solution in January 1959. It was known as 'the monolithic idea'.

Similar to Edison's high-resistance bulb, the monolithic idea was simple but not obvious. Kilby thought that resistors, capacitors and indeed all the components, including the transistor, could be designed and placed on the same monolithic block or 'chip' of silicon. These integrated circuits – or 'microchips' as they became known – would have all the components of a circuit etched onto a single piece of silicon.

Robert Noyce took the idea further by placing a layer of silicon oxide on top of the components like icing on a cake. Tiny lines of a conducting material such as copper could then be printed on top. The problem of having millions of wires connecting the components was therefore solved by removing the wires (Reid 1985). The headache, in this case, was solved by removing the head.

With the monolithic idea, manufacturing components onto big bits of silicon wafer only for them to be popped out and wired into a circuit by the 'transistor girls' on the factory floor, was no longer necessary. The whole circuit would be designed and manufactured without the need for massive wires or the titanic hands of a human working on a production line.

Applications

Microchips took advantage of silicon's semiconducting properties; they had the resilience of the transistor – itself by far the most common component in microchips; they were mass manufacturable; and, because all the components were on the same piece of silicon that were connected by a printed conductor, they did away with the tyranny of numbers by doing away with the pesky connections between each component. Jules Verne's vision was surely about to be made manifest. Almost. Like the telegraph and telephone before it, the microchip was ridiculed and rejected. The manufacturing process was too error prone, the best resistors and transistors at the time were not made from semiconductors and, maybe most importantly of all, the microchip would put most of the world's circuit designers out of a job (Kilby 2000).

Like the telegraph and the battery that came before it, the microchip was also a solution looking for a problem. Like the telegraph and battery, it soon found one. Resistance to the microchip melted away in the 1960s when a genuine need arose. Like Edison's carbon button, microchips could easily be fitted into tiny spaces. They were therefore a perfect match for the Apollo moon mission spaceships and Minutemen ballistic missiles, both of which needed guidance computers that were small enough to fit into their nose cones.

In the first half of the 1960s, the Apollo space program became the largest consumer of microchips. At the same time, the far-sighted Texas Instruments, where Jack Kilby worked, went looking for, and found, another practical application for the microchip. Texas Instruments developed the world's first portable calculators, thus shrinking down earlier mechanical computing devices and, finally, freeing humanity from the mind-numbing number crunching that the arrival of the vacuum tube had promised 50 years earlier.

Where to now?

Despite the fact that the transistor came out of Bell Labs, the telecommunication and computing communities were separate. Nobody had found a reason to bring these distinct technologies and groups of people together.

Then, in 1945, the history of the world once again turned on the axle of humanity's ingenuity. When it did, a truly horrific reason presented itself. Humanity's response to this new terror would be the final link in a chain that bound the magical 1880s, with their generators and light bulbs at one end, and the magical Noughties, with their smartphones and cloud computers at the other. It is to this link that we now turn.

Chapter 6

The wasp and the fig tree

Anyone born in the age of electricity can scarcely imagine a time before the grid. In much the same way, anyone born after the late 1980s can scarcely imagine a world in which communication systems were not digital and widely distributed across optical cables, satellites, through wireless routers and the supercomputers in our pockets that we call phones.

As natural as this fusion may seem to us, computers and telecommunications technologies were once distinct from one another. Their creators, from different scientific communities, saw no reason to bring these isolated technologies together. Then, in 1945, Robert Oppenheimer detonated a nuclear bomb in the New Mexico desert.

Shortly after, a terror arose in the minds of those who had armed themselves with nuclear weapons. It was only a matter of time before one of them was fired. When that happened, a chain reaction would culminate in the deaths of hundreds of millions of people (Ryan 2010). This sent humanity on a mad scramble to combine two godlike powers, computers and telecommunications, so it could save itself from Oppenheimer's third.

The nuclear problem

In the years after the Second World War, America, Russia, the UK and France developed nuclear weapons and playbooks that covered the scenarios in which they could be used (Ryan 2010). One such scenario put my home town in the crosshairs of the Russian military.

Hull is an important but not vital city. It's also on the coast, only 60 miles from where Dracula landed in his ship, the *Demeter*. This meant, in the event of a nuclear strike, depending on the direction of the wind, the fallout would blow out to sea or settle over the Humber Estuary. This, it has to be said, was not general knowledge. This knowledge was instead specific to me and my eight-year-old mates who used to knock about in the tenfoots (alleyways) at the back of our houses. It came directly from the Foreign Office via our mate Andy, whose brother was in the Royal Navy.

Our impending doom was not as far fetched as we first thought. Andy most likely read about it in a public information pamphlet that was published by Hull City Council in 1984. Entitled 'Hull and the Bomb', the pamphlet's cover shows a nuclear explosion in the town centre. The first page begins as follows:

> DEDICATED TO THE VICTIMS OF HIROSHIMA
> AND NAGASAKI
> Perhaps the greatest issue facing mankind today is
> the question of nuclear weapons. No one actually knows
> what would happen if there was a nuclear war.

The pamphlet goes on to say that a nuclear bomb dropped in the city centre would leave a crater that was half a mile wide. The waters from the rivers Hull and Humber would flood into it. Only 6,000 of Hull's population of 268,000 would survive the initial blast. Every building within a three-mile radius of the city centre would be destroyed. That meant my house, our tenfoots, my Commodore 64 and Greenwood Avenue Library would be destroyed, too (Johnson 2022).

According to the government playbooks, if Hull was bombed, then a similar-sized Russian city would be targeted. As long as neither capital – Moscow or London – were attacked, governments could keep talking to each other and therefore discuss terms for a ceasefire or surrender (Ryan 2010). In other scenarios, London or Moscow fell. In that case, each country retained the ability to strike back through its residual forces, such as submarines or bombers on the way to their targets. These scenarios depended on governments and their militaries' ability to coordinate attacks and conduct ceasefire negotiations. Herein lay their problem.

Gamma rays from nuclear explosions in the ionosphere knock electrons out of atoms, which disables radio communications. Point-to-point telephone lines such as the one strung between New York and San Francisco stood no chance of surviving a nuclear blast on land. In other words, in the event of a nuclear conflict, the coordination of troop movements would collapse. Residual forces wouldn't know whether or not to attack. An unauthorised and accidental attack would lead to further escalation. An accidental standing down would pass the advantage to the enemy. It's hard to imagine how this intolerable situation could have worsened. Then, because of Robert Noyce and Jack Kilby, it worsened.

In the 1950s, microchips collided with advances in fuel technology. From that moment on, the US Air Force's Minuteman missiles could be launched in minutes instead of hours. Once the Minutemen were flying, even if launched by accident, every government would follow their playbook, all of which relied on communications systems that would fail instantly. Hundreds of millions of people would die, and most likely by accident (Ryan 2010).

A network that learned from failures

This terror kept a man called Paul Baran awake at night. Baran was not interested in playbooks or the further development of nuclear weapons. He rightly thought a nuclear-proof communications system was the key to averting a disaster.

Baran then came up with an idea for a *decentralised communication network*. Messages would not be sent from one point in the network to another, as they were in the telegraph and telephone systems.

Messages would instead be tagged with a 'to' and a 'from' label. The network then figured out the best way to route the message from the sender to the receiver (Baran 2002).

This routing would be done at relay stations that measured how long it took a message to get to it. The relay stations then shuffled the messages along the quickest route. If parts of the network were damaged by a nuclear strike, the network re-routed messages along the quickest alternative route to the destination. This feature of Baran's network meant that cheaper and shoddy networking infrastructure would easily outperform the best, and most expensive, equipment used by the telephone system.

Packets

As well as learning from its failures, Baran's system included an idea that came to be known as packeting (Baran 2002). Messages chopped into packets and given a unique sequence number made their way along different routes across the network. At the destination, the packets would be reassembled in the right order.

Imagine that, as a kid, I had to get a message from Greenwood Avenue Library, where I picked up that book on BASIC, to the barracks on the corner of Endike Lane and Beverley Road, where there used to be an ack-ack gun during the war. I was only allowed to go straight down Greenwood Avenue and turn right onto Beverley Road. If that path was blown up, my message and I were not getting through. This is exactly what would have happened if a bomb was dropped on Morse's Washington DC to Boston telegraph line.

In Baran's system, the message gets tagged with the recipient, in this case the barracks, and then gets split into ten packets. This time around, I take one packet, Andy takes another and eight of our friends from the tenfoots take the rest. The kids can travel any way they like, including through 7th, 21st and 5th Avenue, through the tenfoots and snickets and even along the fields where our childhood dens were.

At points along their routes, there are desks manned by the kids' parents. These are the relay stations. Each parent makes a note of the time it took for the message to get from the last station to this one (they do this by using a counter stored in the message that Baran called the *handover number*). The parents add that information to the message itself so other relay stations can make similar calculations. The 'intelligence' of the system is calculated at the relay station but forms part of the message itself. In this way, the network learns the quickest route to get packets through and stores that information in the actual message. Depending on the network traffic, those times will change and therefore some packets will go along one route and others will go along another. This means all packets will get there as fast as possible using multiple but different routes.

If part of the network is damaged, the kid with the message backtracks and finds the next available relay station. The parents recalculate the times between the nodes that still exist. As long as there's at least one path from the library to the barracks, all ten packets will eventually make it through.

Later, at the barracks, over a period of ten minutes, the kids straggle in. Some of them are more tired than others. They went the long way round. The last packet arrives first. Then the fifth. The first packet comes in third. Eventually, all the packets make it and the sergeant at the barracks reassembles the message and finally reads it out loud.

Computers meet telecommunications

In real life, Baran's relay stations were never going to be manned by parents from north Hull. In his vision, there would be tens of thousands of these relay stations, which meant they were impossible to physically man. This is why Baran wanted those stations to be manned not by people but by electronic computers.

This was Baran's stroke of genius. He imagined a system that combined telecommunications and computers, fields so distinct that he couldn't find anyone who had experience in both to help him develop his ideas (Ryan 2010). From that combination, an idea for a network emerged that changed its behaviour not just in case of damage from nuclear weapons but also based on network traffic and accidental failures. Distributed across the network and stored in the messages it carried was the 'intelligence' that made it highly adaptive. These ideas eventually coalesced into a system that we call the internet.

Where did they find Paul Baran and which department did they stick him in?

Ever since Edison's invention factory at Menlo Park, new ways of organising knowledge workers had been in the air. Low bureaucracy, flat hierarchies, multidisciplinary teams and learning through failure, all features of modern, humanistic teams, started to appear inside the highly creative walls of places like Bell Labs. The Second World War then forced the military into organising multidisciplinary research teams such as the one that had developed the nuclear bomb under Robert Oppenheimer's guidance out in Los Alamos.

After foundational research won the war, the US Department of Defense (DoD), often through gritted teeth, concluded that groups of people containing a high percentage of mavericks, which were loosely organised behind a vision and under the guidance of a 'conductor' like Oppenheimer, were conducive to using existing technologies to invent new ones.

These dual conclusions, that strict hierarchies did nothing to help create the sort of environment that led to technical breakthroughs and that foundational research won wars, were formalised when the US government birthed a brand new type of organisation. One of them was RAND. That's where Paul Baran worked with 'remarkable' freedom (Baran 2002).

The Intergalactic Computer Network

Another one was ARPA. Like Research and Development (RAND), the Advanced Research Projects Agency (ARPA) was well funded. Like RAND, ARPA had a wide remit to pursue basic, long-term research. Unlike RAND, ARPA was designed to be nimble. The decentralised work of ARPA also had a hidden, political explanation. It was formed during the height of McCarthy's witch hunts. Any centralised government agency at that time invited accusations of communism (O'Mara 2020).

ARPA achieved its nimbleness by teaming up with other organisations such as universities and commercial suppliers to help them carry out their research. This allowed ARPA to disseminate its knowledge widely while taking advantage of the best teams around the United States. Since the ability to direct surviving forces after a nuclear strike depended on the resilience of command and control systems, such systems were at the top of ARPA's to-do list. The man brought in to sort this out for the DoD was J C R Licklider.

Command and control meant only one thing to Licklider. Command and control meant interactive computers.

Human–computer symbiosis

In 1960, well before the use of microchips exploded and more than a decade before the idea of a personal computer was imaginable let alone feasible, Licklider authored a paper called 'Man-Computer Symbiosis' (Licklider 1960). That paper begins with a metaphor. The larva of a type of wasp, the *Blastophaga grossorun*, lives in the ovary of a fig tree. The insect cannot eat without 'the tree and the tree cannot reproduce without the insect'. Together they form a thriving partnership (Licklider 1960).

Machines and humans, Licklider thought, were already like that. Human ability was enhanced by the motor car, the use of the telephone and the strumming of a guitar, all of which provided real-time feedback as the thrum of the engine, the dial tone of the phone or the plucking of a string. The symbiosis brings the machine to life, which in turn bestows power on the human to move at speed, transmit their voice across a continent or make music.

When he wrote 'Man-Computer Symbiosis', Licklider was thinking about computers and humans. If they too could work in symbiosis, a human could not only be made stronger and faster but smarter, too. What if, Licklider thought, a human could 'talk' to a computer, which in turn replied with sensible answers? The human and computer could, together, organise the payroll, work out tonight's dinner, schedule next semester's lectures or even order troops and weapons around the theatre of battle. This is why, when Licklider thought about command and control, he thought about human–computer symbiosis.

The antecedent of the personal computer

The antecedent to the modern, wired-up computer that I'm writing this book on is not really the big, monolithic calculating machines that used to take days to process their inputs into something meaningful. The antecedent of the modern, wired-up computer that I'm writing this book on is radar.

Radar is an acronym. It stands for **RA**dio Detection **A**nd **R**anging. Radar systems send out signals that reflect back off objects such as planes or rain clouds. The reflected signal, whose frequency may have changed depending on the speed of the object it bounced back off, is enough information for the machinery to work out the object's location and velocity. This information, in Licklider's day, was transformed into a blast of electrons that, for a moment, lit up the radar screen. Radar was, in other words, a computing machine that worked and visualised its results in real time. It was therefore nothing like ENIAC.

Licklider realised that if radar was part of a broader command-and-control system, then more elaborate gadgets, working in real time, could be used to win battles by tracking and coordinating troop movements and their supply chains, and by sharing information with the commanders in the theatre just like information on enemy planes was shared on the radar screen (Hafner 1996).

A personal computer and how it could be made to work in a human–computer symbiosis, Licklider thought, had to be the backbone of any computerised command-and-control system. In 1960, however, not only were most of the components for such a computer not yet invented, but they, like the personal computer itself, had not even been dreamed up. What would these components be and how would they be assembled?

The vision

Licklider thought the interactive computer would need a way to share programs and data with other, possibly incompatible, computers. It would have to have a graphical interface, like radar did. It would have to have some way of easily inputting data, too, like a keyboard.

When Licklider arrived at ARPA, different research groups around the United States were working on various parts of this vision. None of them were working on the whole thing. In fact, none of them had even envisioned the whole thing. Licklider's job was to plant this vision in their minds while battering these diverse groups into a community.

The way to do that, at first, was to get them together and, over drinks and late into the night, swap ideas as Licklider's vision took hold of their minds. This is why he was always on aeroplanes and not at his office at the Pentagon, where the sign on the door read:

ADVANCED RESEARCH PROJECTS AGENCY
Command and Control Research
J C R Licklider, Director

One day, 25 April 1963, in that very office, just before dashing off to get on another plane, Licklider hastily dictated a memo (Waldrop 2018). Its heading read:

ADVANCED RESEARCH PROJECTS AGENCY
Washington, DC, April 23, 1963
MEMORANDUM FOR: Members and Affiliates of the
Intergalactic Computer Network
FROM: J C R Licklider
SUBJECT: Topics for Discussion at the Forthcoming Meeting

He had never called the group the 'Intergalactic Computer Network' before. By doing so, he was making it clear that they were a group. As to the memo itself, by Licklider's own admission, it was rambling. But it held within it the seeds of the future. He made clear that the groups had to build on each other's work and that their programs and data must be easy to share over some sort of (yet-to-be-invented) network. Smack bang in the middle of the memo, Licklider said, 'If such a network as I envisage nebulously could be brought into operation, we would have at least four large computers, perhaps six or eight small computers, and a great assortment of disc files and magnetic tape units not to mention the remote consoles and teletype stations – all churning away' (Licklider 1963).

That sentence is crucial. Licklider was proposing a system for members of the Intergalactic Computer Network to communicate with each other. To share. To collaborate.

The communication system that the Intergalactic Computer Network would go on to build for themselves – the internet – would be the nuclear-proof command-and-control system that the Pentagon wanted and that Paul Baran had, at least partly, envisaged.

When he wrote that memo, did Licklider realise the system he proposed was the end product, the central focus of the Intergalactic Computer Network's research and the way they would run and share the results of experiments? (Waldrop 2018) I think he did. Licklider knew that it wasn't enough to invent the future. You had to live in it, too, in the exact same way that Thomas Edison had once lived in the future out at Menlo Park.

The ramifications

Licklider's contribution cannot be overstated. He provided a vision and managed to get the whole of the Intergalactic Network to buy into it. It was in this way that the idea of the interactive and networked computer jumped from his head into the heads of those who had to build it. However, as well as injecting the human–computer symbiosis virus into ARPA, he also injected the gigantic amounts of cash that he had in his budget, which he allocated to groups that contributed to his vision.

Within just a couple of years, Licklider created a vision, brought together a range of research teams into his Intergalactic Computer Network and by doing so left ARPA enthused about what was possible. In doing all this, we now know Licklider set in motion a chain of events that would give humanity the personal computer, the mouse and the graphical user interfaces needed to interact with it, the internet itself… and therefore cloud computing. Licklider also gave humanity and computing another important Bob. This one was called Taylor.

Another son of a preacher

Bob Taylor liked responsive machines. He learned to drive when he was 11. It never crossed his mind, however, that a computer could also be used responsively (Waldrop 2018). This changed when he read an article in 1960 about interactive computers by one of the most famous researchers in Taylor's own field of psychoacoustics: J C R Licklider.

'Man–Computer Symbiosis' inspired Taylor to move to NASA, where he was put in charge of a program that funded research into flight control systems and computers (Waldrop 2018). He gladly used some of that money to move the state of interactive computers forward, like when he funded a study of computer-assisted air traffic control.

When Licklider and Taylor finally met in 1962, they got on well.

They were both Southerners and, like Bob Noyce and Samuel Morse, were both sons of preachers (Smith & Alexander 1999). Taylor was so passionate about the interactive computer, and made such an impression on him, that Licklider brought him to ARPA in 1965. Taylor became deputy to Licklider's successor, Ivan Sutherland. When Sutherland moved on, Taylor took the job and thus became the third director of command-and-control research at ARPA. By then, though, since Licklider had renamed the group in 1964, it was called IPTO – the Information Processing Techniques Office.

What came first, the network or the personal computer?

When he took that job, two ideas were percolating in Bob Taylor's mind. Communities of real users had formed around computers at places like the University of California at Berkeley. This pleasantly surprised and then fascinated Taylor. The more he thought about it, the more he believed that communications and community would be central to the future of computing.

That was the first idea Taylor was brewing. The interactive computer was the other. It was the interactive computer, after all, that would allow humans and computers to be coupled together in a creative partnership, as Licklider had written in 'Man-Computer Symbiosis'.

These two ideas collided in Taylor's mind. Why not have a great network and use it to connect interactive computers and their communities together? It may seem obvious to us but in 1965 Bob Taylor's idea was part genius, part heresy and part lunacy.

ARPANET

Taylor decided that the network might as well come first. Early experiments had moved bits over bog-standard telephone lines in the same way that they'd once gone over telegraph wires as dots and dashes. Why not try to change Licklider's Intergalactic Network, which already was a community, into a wired-up community? Taylor almost certainly knew that one day, once the network was up and running, the Intergalactic Computer Network's clunky computers could be replaced by true, interactive computers (Hafner & Lyon 1996).

Of course, ARPA would have to invent them first, which was part of Taylor's decision-making process. Networking was a known quantity and well within grasp. Interactive computers, on the other hand, were well and truly in the future. So far into the future, in fact, that they had not fully emerged in the imaginations of those who would go on to invent them.

Mission accomplished

Bob Taylor succeeded in launching the ARPANET and as soon as he did, he left. During his time as director, Taylor, unlike most other wartime researchers, spent time in the theatre of war. His rank of brigadier general ensured protection under the Geneva Convention. Taylor needed that for his trips to Saigon, where he worked on a computer-based reporting tool. It helped the military to have one view of the situation there. Taylor said it was all lies but, after he helped, they were at least consistent lies.

It bothered Taylor that ARPA was part of the Department of Defense. He comforted himself with the knowledge that the work they were doing might one day have a profound impact on society. His work was, after all, relatively benign – although the introduction of Agent Orange to the war in Vietnam was also an ARPA project. That, of course, doesn't fit ARPA's creation myth. ARPA may like to boast about its willingness to fail but, as Weinberger said, 'That does not mean that it is eager to have those failures examined' (2017). Jack Kilby and Bob Noyce no doubt thought their work was benign, too. They had no idea that, when coupled with advances in fuel technology, their microchips would enable push-button genocide. We don't know if Taylor knew Agent Orange came from ARPA.

What we do know is the war made him sick (Waldrop 2018; Hiltzik 2000).

Taylor's growing disillusionment over the war was already on the top of his mind. Richard Nixon's victory in the election of 1968 pushed him over the edge. He had had enough. His former boss and predecessor, Ivan Sutherland, was at the University of Utah. Taylor asked if he could join him there. Sutherland said yes.

By the time Bob Taylor had grown sick of the war and the DoD, the interactive computer, that beautiful virus that was so expertly cultured and spread by Licklider, had outgrown him, outgrown Taylor and completely outgrown the military. Modern Prometheans had once again raided Mount Olympus, only this time they had not brought back Zeus's thunderbolts. They had brought back something else entirely.

End of Part 2

The control of electricity evolved from Volta's batteries to the telegraph, Edison's light bulbs and their progeny, the vacuum tube and later the transistor and the microchip. The switch-on-and-off-ability of the vacuum tube allowed it to be used in computing machines. However, their limitations were soon revealed. Computers made from hacked light bulbs melted the machinery they were part of, had massive energy needs and broke down constantly. Unless they could be miniaturised, Jules Verne's vision would be forever out of reach.

Things changed with the arrival of the fabulous midget in 1947. The 'magical' transistor kicked off miniaturisation and after that computing and computers became more 'magical' too. But miniaturisation got off to a slow start. The problem was not that electronic components could not be made very small with silicon. They could be. The problem was that these components could not be wired together. In the beginning, their assembly, as evidenced by the innards of the transistor radio, was done by hand.

The monolithic idea, that all the components of a circuit could be gathered on one piece of silicon, not only solved the tyranny of numbers but converted the assembly of circuits from a manual process involving gigantic human hands into a manufacturing process.

Microchips had an almost instant impact on humankind. One thread of activity, the electronic calculator, headed by Kilby at Texas Instruments, led to the eventual freeing of human computers and clerks from the mind-numbing number crunching that made up the majority of their work.

Another thread, however, saw guidance computers added to

ballistic missiles, after which nuclear conflict could be started, possibly by accident, in minutes instead of hours. It was the terror of nuclear annihilation and the government's woefully inadequate command-and-control systems that led Paul Baran to dream up a nuclear-proof communications system, while J C R Licklider developed his ideas of human–computer symbiosis. These ideas, at first through the work of ARPA, developed until they allowed millions of customers to browse an online catalogue of books at Amazon. In 1969, however, that was well and truly in the future.

What was not well and truly in the future was the digital computer. Digital computers started off life as lots of hacked light bulbs in machines like ENIAC, which needed constant debugging. Not long after that, though, microchips and their split-second number crunching arrived. They did that number crunching inside the nose cone of ballistic missiles or in the innards of desktop calculators. The world had turned on the ingenuity of Robert Noyce and Jack Kilby. It would not be turned back.

Themes redux

We have once again seen that new technologies are often solutions looking for problems. This was true for the vacuum tube, true for the transistor and true for the microchip.

Related to this, we again saw that a technology's destiny is shaped not by its creators but its users. Walter Brattain and John Bardeen had no idea what Noyce and Kilby would do with the transistor, who in turn had no idea that their invention would almost instantly become a tool of push-button genocide.

We have also learned one extra thing about systems. In Part 1 we saw that a change in just one component, the light bulb, can change the nature of an entire system. However, in Part 2, we saw that a change in one component can leave a system intact but alter its performance beyond all recognition. The transistor did nothing to the overall design of the radio. What it did instead was to make it cheaper and lowered its power needs so that it could run on a battery not dissimilar in its power to one of Alessandro Volta's piles.

This explains why the transistor radio was cheaper ($49.95, when it was first launched in the US), better quality, smaller than any radio ever produced and was thus living proof that the transistor eliminated all the problems with vacuum tubes in one fell swoop (Reid 1985; Gordon 2016). The portability of the transistor radio also afforded privacy; teenagers listened to something new called 'rock and roll' away from the judgemental ears of their parents (Reid 1985).

We also learned something new about computers, caught a glimpse at their shapeshifting nature and learned a little more about the transformability of information.

In terms of shapeshifting, we saw that the computer was used both for generating firing tables and as a real-time tool (radar) for tracking enemy movements. The computer's ability to shift shape like this is why writing its history is so difficult. What is a computer and what is it not?

With regard to the transformability of information, in Part 1 we saw that letters can be transformed into dots and dashes. In Part 2, we saw that computers could transform information about planes into firing tables. In the 1940s, researchers wondered, is there anything information cannot be transformed into?

Finally, we saw that the impact of humanistic management was so great that even the Department of Defense understood the effectiveness of creating space so that mavericks could be free to do their mavericking.

Where to now?

As the 1970s dawned, advances in management, microprocessor technology and software development were about to collide. These titanic forces did not collide in a bookstore. They instead collided at a photocopying company. The interactive computer's time was about to come.

Part 3

THOSE WHO KNEW THE SCORE

Chapter 7

The Alto and the internet

Xerox built its whole business around one technology, xerography. Xerography is Greek for 'dry writing'. The modern Prometheans at the photocopying company didn't steal fire or thunder but instead raided Mount Olympus and nicked... Hermes's pens and paper?

Xerography was a revolution. In exactly the same way that people didn't vacuum but instead *hoovered*, didn't search the internet but instead *googled*, back in the olden days, office workers didn't copy but instead *xeroxed*. With sales of more than a billion dollars, in the 1960s, Xerox was a fantastic business. But it was not a daft business.

Pioneering companies such as Lyons in England demonstrated that the future of the office was computerised. Xerox quickly came to see that their future would be computerised, too. They had the vision for an 'electronic office' and importantly the financial resources to invent it. Soon, a search began for the best person to head up the computer division of Xerox's newly minted Palo Alto Research Centre (PARC). Over and over again, that search returned the same name: Bob Taylor.

The timing seemed perfect, not least because one year in Mormon country was just about all Taylor could handle. When Xerox came calling, it started to look like Bob Taylor was going to get to make computing history. Again.

Licklider's kids

In his remarkable book, *The Dream Machine*, M Mitchell Waldrop calls the formation of Palo Alto Research Centre (PARC) a near-miraculous, never reproducible coming together of talent, luck and timing (2018).

There were a few reasons for this. First of all, because they got Bob Taylor, the people, ideas and the living-in-the-future culture of ARPA was transplanted to PARC. PARC was, in other words, a direct continuation of the work that had started at ARPA (Waldrop 2018).

Second, the most powerful work is never done by the adults who invent a technology but by the first generation of kids who grow up with it. We saw that in the 2010s when my generation went on to do the most innovative work with the cloud. The generation behind mine will almost certainly make the biggest contributions to artificial intelligence.

Similarly, it was Licklider's 'kids', whom he had brought together and whose research he had funded through ARPA, who were assembled at PARC at the precise moment they were coming of age. Under Bob Taylor's guidance, they would do the most important work of their lives (Waldrop 2018).

Finally, like AT&T before them, Xerox had mountains of research dollars to spend. Those dollars came with a promise to Taylor that he would be left alone. This miraculous coming together meant that, after a year decompressing in Utah, Bob Taylor was going to get the chance to invent the interactive personal computer and stick a load of them on the same sort of network that he had dreamed up for ARPA.

The Alto

In the beginning, Taylor and his newly assembled team had no idea what the personal, networked computer was going to look like. By 1972, that had changed. The team had more or less worked out their prototype, which they called the Alto. Its features might sound familiar to you (Ryan 2010; Waldrop 2018).

- The Alto would have a screen about the size of a piece of paper. It would be white and the letters beamed onto it would be black. In other words, the Alto would look like the word processor I'm writing this book on. Since the machine was gestating at Xerox, this was maybe not surprising.
- The Alto would have a keyboard and a funny thing to control the cursor that Taylor's team called a 'mouse'. The mouse manipulated 'windows' on a graphical user interface.
- The Alto would communicate with other computers on the network. In the early 1970s, computers were still mind-blowingly expensive, while not being that powerful. If the machine had too much work to do, and if it was to remain fast enough to be interactive, it would have to offload data and its programs to other machines to pick up the extra load.
- Finally, there was going to be a cabinet – not the size of a filing cabinet, but something more like a foot locker that would sit under the desk.

By 1972, the team at PARC had, in other words, specified a system that looked like the personal computer as we know it today. All they had to do was build it. Could it be done? Possibly.

Semiconductors come of age

When he was still in Utah, Bob Taylor did some maths. If Jack Kilby and Bob Noyce's microchips kept improving at the same rate, soon enough the personal computers that Taylor hoped to build would be powerful enough to be useful and cheap enough to be practical. That started to happen in the early 1970s. Therefore, at the exact moment that Licklider's kids were coming of age, so too were microchips.

If his maths and Gordon Moore's law (that the number of transistors that could be crammed onto a chip would double every 18 months) held throughout the 1970s, then Bob Taylor's job at PARC would be to provide the vision, the leadership and the management. He would keep the team stocked up on Dr Pepper, which he stored by the pallet load in a secure locker, and he would keep the suits off their backs like he had once done at ARPA. While all that was happening, advances in microchips, he hoped, would make the machinery PARC was inventing cheaper and faster.

Completing Baran's vision

Meanwhile, ARPA was not stationary. Not only was it implementing Taylor's previous vision, it was also improving it. Taylor's ARPANET was about wiring machines together using physical network connections. It did not connect satellites and radio transmitters in the way that Baran had envisioned. In fact, nobody had a clue how radio could be used to network computers together.

That changed in 1970. Funded by ARPA, Norman Abramson, a professor of electronic engineering and computer science at the University of Hawaii, wanted to link the university campuses' computers. The problem was that they were on different islands and the telephone network was shoddy.

Abramson decided to use radio. The problem was that messages broadcast by different nodes on the network at the same time became garbled. Abramson's solution was simple. If packets collided, the nodes waited a random amount of time before trying to send them again. The chance of collision the second time around was lower. However, the chance of further collision tended towards zero as each node waited a random amount of time before trying to get its packets through again. This packet radio network was called AlohaNet and its technique for waiting became known as the Aloha System (Ryan 2010).

Two more Bobs

Norm Abramson's ideas found their way into Xerox PARC because of two more important Bobs. Bob Metcalfe worked on the ARPANET for Licklider while he was doing his PhD at Harvard. Before he graduated, he accepted a job offer from Bob Taylor out in California at PARC. He was, however, not free of his duties for ARPA. He was still getting on and off aeroplanes as he helped organisations get onto the ARPANET. The jet lag caused havoc with his sleep patterns. To knock himself out one night, he picked up Abramson's 'Aloha System' paper. The paper had the opposite of the desired effect. Metcalfe was riveted.

Like Edison's carbon button, AlohaNet was a beautiful design. It meant each computer on the network was totally free of the others and of a central coordinating computer. It was, in other words, a step towards Baran's decentralised network. Metcalfe realised that the medium did not matter. Whether the network was something in the air over the seas of Hawaii or a physical network in an office, chucking packets into the ether as the machines followed a shared protocol was a great idea. Even if they collided, they would eventually make it to their destination (Ryan 2010, Waldrop 2018).

Not long after that sleepless night with Abramson's paper, Bob Metcalfe typed up his notes in a top-secret Xerox memo. The original name, the Alto Aloha Network, vanished. It was replaced by ETHER Network. Metcalfe had realised that the Aloha Method was a good idea even for a physical network because it removed the central node and left the coordination of sending the message to the computers following the protocol.

Not long after that, Metcalfe teamed up with David Boggs. He was part time at PARC as he finished up his PhD at Stanford. Across the campus from Boggs, Bob Kahn from ARPA nudged his colleagues, including Vint Cerf, to organise lectures on how packets could flow between the ARPANET and other networks like AlohaNet. More generally, packet networks were appearing everywhere and Bob Kahn wanted to work out how these networks could be networked together into one, giant, *internetwork*.

So, who invented this internetwork or 'internet' for short? Packets came from Baran but their logical simplicity meant they were simultaneously invented in more than one place. TCP/IP came from ARPA as part of the work kicked off by Taylor but he was long gone when that happened. Kahn himself said he has no idea where the word 'internet' came from or exactly when it started getting used. Regarding the internet, Waldrop said there was just something in the air (2018). They were the exact words that Noyce had used to discuss the microchip's invention. Noyce felt he could not take credit for it because in 1958 and 1959 there was just something 'in the air' (Berlin 2005). The internet and chip, in other words, emerged from the collision of ideas launched by multiple people in multiple places at more or less the same time.

The last bits

Bob Metcalfe went on to perfect Ethernet at PARC. ARPA built their own packet radio network – PRNET – based on Abramson's AlohaNet (Ryan 2010). In 1975, ARPA initiated the SATNET programme using Intelsat IV Earth stations in the US and UK. From that point on, all access to the ARPANET from the UK went through SATNET (Ryan 2010).

In 1975, there were therefore three working packet networks, ARPANET, SATNET and PRNET (Ryan 2010). These three were then inter-networked using protocols that Vint Cerf and Bob Kahn went on to invent after their lectures in 1973. They were called Transmission Control Protocol and Internet Protocol. They are known as TCP/IP.

In 1979, tests showed that the internet worked flawlessly, with packets always eventually getting through, even in tests that simulated a nuclear attack. Paul Baran's vision had (finally) been brought to life.

The virus escapes

During those same years, Taylor's team invented the personal, interactive, and because of Metcalfe and Boggs, *networked* computer. While they were at it, they invented the laser printer, the word processor and, of course, Ethernet, which had started life as

AlohaNet. The team at PARC really did invent the office of the future.

Along the way, though, the PARC virus was released into the wild. Computing myth says that Steve Jobs and Bill Gates came along and stole PARC's ideas and then built their own computers (in Apple's case) and operating systems (in Microsoft's). It's a good story, but it isn't true (Waldrop 2018; Hiltzik 2000).

The inventor of the word processor at PARC, Charles Simonyi, later moved to Microsoft and worked on a product they were developing called Word. Simonyi said that he was 'the messenger RNA of the PARC virus'.

Steve Jobs, on the other hand, caught the bug directly. He was invited to PARC and shown what they were building. He didn't steal their ideas. He copied them with their apparent blessing – a blessing that was later revoked in 1989 when Xerox unsuccessfully sued Apple.

The future blown

Once the PARC virus was out, there was no way it could be reined back in. Xerox had invented the future but before making any money with it – in fact before the work had been completed – it escaped. This was arguably the biggest blunder in the history of business.

How did Xerox invent the future and not reserve the right to play in it? One reason is because Xerox was an East Coast company and it clashed with the West Coast hippies of PARC and their way of doing things. Xerox did not catch the PARC virus that they created. Instead, its immune response killed it.

Part of that immune response was to Taylor himself. They hated him and they hated the culture he created at PARC. This stemmed from their poor understanding of the computer and their even poorer understanding of how to manage the mavericks who invented it. In 1972, the journalist Stewart Brand visited PARC. He wrote an article for *Rolling Stone* called 'Spacewar, Fantastic Life and Death Among Computer Bums'. Xerox was apoplectic. A photograph of Taylor smoking his pipe appeared on page 4. The implication of the article was that tobacco was not the only thing being smoked at PARC.

Apart from Taylor and his computer bums, Xerox had another problem. They had no idea what to do with the personal computer. They eventually built an electronic office based on the Alto called 'Star'. Xerox Star could print and send messages to other computers on the network but unfathomably, it did not include Charles Simonyi's word processor, the one thing that every secretary and every office of the future needed.

When Star eventually came to market in 1981, it was sold as an entire system of computers, file servers and high-quality printers. It cost, in today's money, more than $50,000. The problem for the Star was that Taylor's maths was proven to be correct. In the eight years between the invention of the Alto and the release of Star, personal computers and the computer programs that ran on them became cheap and useful. The Apple II was a stand-alone machine. It sat on the desks of executives and secretaries like the Commodore 64 sat on mine. Given the economic hard times of the early 1980s, why would any business pay $50,000 for a computer when something just as good was on the market for a couple of thousand?

While it attacked itself, the future sailed right past Xerox. In 1981, it was a worn-out copy of its 1971 self (Smith & Alexander 1999). It never recovered.

The score and the conductor

The computer was an important part of the story at PARC. But so too was Taylor's management. At PARC, like Edison at Menlo Park, Taylor had to find a balance between directing his people and letting them get on with their jobs. Back then, there was no management theory that explained this. A theory that explained PARC arrived in 1993.

In *Post-Capitalist Society* (1993), Peter Drucker provided the best explanation as to why teams like the one at PARC worked. Drucker said that teams of knowledge workers are organised like a symphony. Those in a symphony have clear roles. The tuba player will not pick up the violin but both musicians do work in relationship to each other. That work is coordinated by a conductor. At PARC, Bob Metcalfe worked on Ethernet and Charles Simonyi worked on word processors. Their distinct but clearly related work was coordinated by Taylor in the role of conductor.

In a symphony, the orchestra and the conductor follow the same score. The personal, distributed computer was the 'score' that all the 'musicians' at PARC followed. There's no need for excessive communication or control hierarchies when everyone is following the same score. The job of writing that score and conducting the orchestra once fell to Edison, Licklider and, of course, Robert Oppenheimer in the desert of New Mexico.

The importance of following the same score was dramatised brilliantly by Christopher Nolan in his 2023 film, *Oppenheimer*. General Groves, played by Matt Damon, is worried about security. He wants Oppenheimer to compartmentalise his teams, making sure that none of them sees the whole picture. Groves, in other words, wanted to limit how much of the 'score' each 'musician' in the 'orchestra' knew. Oppenheimer, played by Cillian Murphy, replies, 'All minds have to see the whole task to contribute efficiently. Poor security may cost us the race; inefficiency will' (Nolan 2023).

After the war, Oppenheimer testified that 'We needed a central laboratory devoted wholly to this purpose, where people could talk freely with each other, where theoretical ideas and experimental findings could affect each other, where the waste and frustration and error of the many compartmentalised experimental studies could be eliminated, where we could begin to come to grips with chemical, metallurgical, engineering and ordnance problems that had so far received no consideration' (Rhodes 2012).

Compartmentalisation of teams can make them easier to manage and certainly easier to conceptualise. But at what cost? Xerox came last in the race that they started.

The musicians matter

Edison, Licklider and Taylor, not to mention Oppenheimer and the leaders at Bell Labs, had another thing in common. They valued talent. They could spot it, nurture it and extract it where there seemed to be none.

Taylor the conductor

Taylor himself took his job as a manager seriously. He nurtured a democratic culture, insisted that it was not enough to invent the future but that his teams had to instead live in it; he removed obstacles and guaranteed an endless supply of Dr Pepper. In his role as conductor, he made sure that everyone in the orchestra followed the same score. This meant that separate but related work streams moved forward together.

Finally, Taylor knew that a system only worked if its components worked well together. The Alto's components worked well together because Bob Taylor got their strong-willed and often egocentric creators to work well together. Edison would have understood, would have known the score.

Where to now?

In the New Mexican desert, Robert Oppenheimer toppled a domino that started a chain of events that culminated, in the late 1970s, with the creation of the personal computer and a nuclear-proof communication system called the internet.

That chain of events, however, had nothing to do with the development of the microchip. The microchip developed in parallel, flirted with the computer along the way, terrified Paul Baran when it was added to the Minuteman missiles, but was nevertheless part of a different chain of events that had started at Bell Labs with the invention of the transistor.

These two parallel but distinct technologies, personal computers and microchips, did not start to come together, at least not properly, until in the mid-1970s. When that happened, the final pieces of the puzzle that allowed Amazon to sell books on the internet were put in place. We now have to return to the summer of 1968 to find out what was happening to the descendants of the ENIAC.

Chapter 8

The rise of utility computing

Once upon a time, in much the same way that big factories could afford their own power generators (even though they'd have preferred to plug into the not-yet-invented grid), big companies could afford to build their own computers before they were commercially available. One such company was J Lyons & Co, which had, by the 1960s, grown from its humble beginnings in 1894 as a single tea shop into a chain of cafes and hotels, including the Trocadero in London's West End. By that time, it was a serious food manufacturer, bringing great comfort to me and my friends in the midst of nuclear annihilation. That great comfort came in the guise of Lyons' Battenberg and Bakewell tarts.

In the late 1940s, two executives from Lyons went on a fact-finding mission to the United States, where they met one of the designers of ENIAC. Upon their return, they continued their mission, visiting Cambridge to see another computer, EDSAC (Electronic Delay Storage Automatic Calculator), which was being developed by the university there.

Impressed by what they saw at Cambridge, Lyons wanted to know how such a machine might one day help them. Because they wanted an answer to this question sooner rather than later, they brought the development of EDSAC to an early completion by providing essential extra funding and an equally essential pair of hands in the form of an electronic engineer.

When it was finished, and after seeing EDSAC running

successfully, Lyons began work on their own machine. They called it LEO – the Lyons Electronic Office (Ferry 2003).

The beginning of commercial programming

LEO was rolled out in 1951 to relentless mockery by the British press. They wanted to know if this tea shop 'brain' would be used to calculate the number of beans that should go on toast or just how much extra Swiss roll would be needed for the school holidays (Lean 2016). This mockery didn't change the fact that LEO did the serious work of calculating tea blends and efficiently automated the ordering of pies and cakes to Lyons' nationwide network of cafes, doing the work of 300 clerks but with annual running costs that were equivalent to just 50.

The mockery also didn't change the fact that Lyons's board of directors were far sighted. They saw the potential of computers in business when hardly any other companies did. In 1952, when Lyons hired the world's first commercial computer programmer, Mary Coombs, the age of commercial programming had begun, of all places, in a tea company.

Punch cards

If you remember Lyons' Battenbergs and your grandma's glowing radio, you may also remember fairground organs, the ones that played themselves. These hungry machines ate reams of perforated or punched cards, transforming the holes into notes that the organ played.

Weavers programmed old looms in the same way. Pattern data, stored on punch cards, was fed into the loom, along with the threads, and intricate tapestries came out the other end. Joseph Marie Jacquard, who patented the Jacquard loom, had a set of cards that produced an image of himself that was so detailed, it looked like a photograph. This might be the earliest example of a high-definition digital image.

Looms and fairground organs were not used to calculate the orbits of planets, like human computers once did, but they transformed data on punch cards into tapestries and music like ENIAC had transformed trajectories into firing tables.

Mainframe programming

LEO, like other mainframes, so called because their parts were stored in large cabinet frames, was programmed in exactly the same way as fairground organs and Victorian looms. Punch cards went in and results came out. The results were not music or a new tapestry but the output of the program printed neatly on slivers of paper. As well as ending with paper, the process of programming also started with it. Back then, programs started life on pads of paper, written out by the programmer. These programs were not readable by a computer, making them *pseudocode*.

When the program looked as if it might work, the programmer translated it into holes on punch cards, thus making it machine readable. The programmer then compiled the cards into the right order, like publishers used to compile manuscripts, and fed them into the computer. The holes controlled the forming or breaking of electrical circuits, translating them into signals that were directed through vacuum tubes or, in later mainframes, Kilby and Noyce's microchips.

Once the circuitry had them, these flows of electrons were manipulated exactly as they were on Nemo's *Nautilus*. The results of these manipulations, when they were ready, were translated back into something a human could read and printed on a bit of paper.

Coding

The manipulation of electricity in this way, as the telegraph already taught us, is nearly always about 'coding' – encoding and decoding. It was no different for mainframes like LEO. Programming a mainframe was about encoding thoughts into pseudocode and then pseudocode into punch cards so that the computer could finally convert the holes in the cards into electrical signals. It was the coding that made these computers digital, like the telegraph was and the telephone was not. On the way back out of the computer, the process was reversed and the signals in the machine were decoded into something a human could understand.

Programmer experience

The program, once it was translated to punch cards, could not contain any errors. Unfortunately, the chance of writing an error-free program on the first attempt and then translating it to the machine-readable language of punch cards was practically zero. What usually came back out of the machine, then, was an error message.

This would not have been so bad if the error was easy to correct. But it took hours and sometimes days after the programmer fed the cards into the computer for the result to be printed out. The error's arrival signalled the beginning of the laborious task of debugging. Once the area of the punch card with the bug was found, the programmer patched it, usually with sticky tape, in the same way a punctured bike tyre is patched. They then fed the patched cards back into the computer and the whole cycle began again. The most likely outcome was a new error. It was a horrendous way to work.

The priesthood

Another feature of mainframes made an already horrendous process even worse. Because they were expensive and delicate, their climates having to be tightly controlled, those in charge of mainframes restricted their access. Programmers therefore handed their punch cards through a slit in the wall to often faceless operators. The results eventually came out on the other end hours or even days later, depending on how many other programs the operator had batched in front of it.

It didn't take long for the operators to form a priesthood that zealously guarded access to their monolithic, electromechanical holy of holies. Recast against their wills as acolytes, 1950s programmers railed against the system (Levy 2010). For them, the power of the computer was within sight, but its access, dictated by the whims of the priesthood, was never fully in reach, while programming and debugging with punch cards was excruciating.

The drudgery of computing hadn't gone away. It had instead transformed into mind-numbing programming, debugging, patching and grovelling to the high priests of the mainframe. The world of computing was therefore primed for, and was fortunately about to go through, a revolution that lined the two biggest bugaboos of mainframe programming up against a wall and shot them both through the head.

The first cloud

In 1968, a supercilious eighth grader returned to Lakeside school after the summer break and found a peculiar machine that had a keyboard, printer and funny roll of yellow tape attached to it (Manes & Andrews 1993). This machine was definitely not a computer. But it definitely wasn't a typewriter, either. It was instead, as Maines and Andrews called it, an 'electromechanical contraption' that held within its workings the remnants of the mechanical and analogue world of Edison and Bell and a glimpse of the informational and digital world that at that moment was being brought into existence (1993). It was, in other words, an evolutionary link between what came before and what would come next. It had, of all places, turned up in a school in Seattle in much the same way the BBC turned up in my school.

As well as the roll of paper and its keyboard, the machine, an ASR-33, had a modem that helped connect it over the phone lines to a distant and hidden General Electric computer (GE-635). The modem's job was to modulate, or change, the ASR-33's digital signal into an analogue signal so it could be carried over the telephone line. On the other end, the analogue signal was demodulated, or changed back, into a digital signal so the remote computer could understand it. Modem is a portmanteau of **mo**dulate and **dem**odulate.

The GE-635 was different not because it was not a mainframe – it was – but because it had the Dartmouth Time-Sharing System (DTSS) installed on it. DTSS was the first really successful large-scale time-sharing system. This remarkable program allowed multiple users to share access to a mainframe without having to be in close physical proximity to it. In other words, time-sharing systems meant that a single computer could be shared by multiple users in multiple places at the same time. This lowered the costs so much that businesses, universities and even high schools and individual users could now afford access to a computer without having to own it.

The computer companies pulled off this magic trick by renting their customers 'dumb terminals', which is what the ASR-33 was. They then charged each customer for the exact amount of computer processing time they used. At the end of the month, each customer received a bill from their computer company in the same way they would get a bill from the electricity company.

Because of the magic of time-sharing, what was previously available to only the richest corporations, well-funded universities, British cake makers and government agencies was, in 1968, made available to school children in Seattle with monies not from a generous grant but from the proceeds of the Lakeside Mothers' Club annual jumble sale (Manes & Andrews 1993). Time-sharing democratised computing, removed the burden of looking after the machine, and changed what was previously a titanic capital cost into a monthly expenditure based on usage. The proto-cloud had arrived.

The triple whammy

Time-sharing computers had three features, each of which corrected problems with mainframes that allowed them, in the blink of an eye, to usurp computers such as LEO, which themselves had only just risen to prominence a few years earlier.

Two of these features dealt directly with the bugaboos of mainframes. The third allowed users to interact with the computers conversationally, as Licklider had envisioned in 'Man-Computer Symbiosis'.

1. Programmer experience

Time-sharing computers did not use punch cards. Instead, users typed their programs directly into the dumb terminal. When it was ready to go, the user sent the program and all its data to the remote computer. When the program was finished, the results were sent back and printed out. In the case of the ASR-33, the results were printed on the yellow ticker tape. It was in this way that computers became interactive; the user sent a command and the computer sent a result back. The conversational nature of this interaction created the illusion that a single user had full and unfettered access to the mainframe. This experience was, in other words, a programmer's dream come true.

2. Operating systems

Time-sharing systems needed a way to provide users access to the mainframe, calculate billing information, secure user data, allocate memory and process incoming messages from, and format outgoing messages to, the dumb terminal. This was achieved with a new type of computer program called an operating system. DTSS was an early and successful example of an operating system. This marked a sea change in the history of computing. From that point on, computers would no longer be managed by people; they would be instead managed by programs.

3. High-level programming languages

DTSS was developed at Dartmouth College in the United States. The motto of the college is 'a voice crying out in the wilderness'. Before time-sharing came along, this was exactly what thousands of clerks and computer programmers had been doing for a decade.

This cry was answered by the team at Dartmouth, who not only developed the operating system but a simple and easy-to-use programming language that they called the Beginners' All-purpose Symbolic Instruction Code (BASIC). To understand how important this change was, consider this simple program, which once it was punched in holes on a card, was almost unreadable:

```
Load B
Add C
Store in A
```

The 'Load' command brings a number from the computer's memory. The 'Add' command adds the value stored in C to it. Finally, the 'Store' command puts a number back into the computer's memory. The same operation in BASIC, however, is much easier to understand:

```
A = B + C.
```

BASIC, a 'high-level' programming language, relieved programmers from the mind-numbing burden of instructing the computer exactly where to grab numbers from and exactly where to store them. With BASIC, the programmer focused on what they wanted the computer to do and the programming language translated that into instructions that the computer understood. 'Compiling' programs was no longer about compiling punch cards together. Compiling was now automated, the hard work being done by a computer program, too (Ceruzzi 2012).

Child's play

The triple whammy of time-sharing made programming a mainframe, a hitherto jealously guarded, multi-million-dollar electronic 'brain', child's play. After their return to school in 1968, the Lakeside kids proved this by mastering the ASR-33, the GE-635 it was connected to and BASIC in a matter of hours.

In those few hours, the kids also become instantly and hopelessly hooked. They spent weeks and weeks in the computer room that academic year. Photographs taken at the time showed a familiar but unbelievable scene. One picture included two high school students, one whose feet don't touch the ground, working on the ASR-33 and therefore working remotely on the GE-635, the sort of machine that was, at that time, inaccessible to all but a small number of clerks, operators, academics and those in the military.

Still too expensive

The Lakeside kids' experience in 1968, compared to mine in the 1980s when I had my own little computer, were almost identical except for one crucial difference. After their BASIC programs were executed and the results returned, the kids dialling into the remote computer in 1968 got one extra piece of information: the amount of time that the computer was busy running their program. This essential information indicated the costs that, at the end of the month, were totalled up and sent as bills based on their usage. If they knew the time it took to run, programmers had an idea of how much their program cost. This allowed them to estimate what their bill would be and, if possible, look for ways of reducing the running time of their programs in order to save money. (Companies all around the world are currently engaged in the exact same sort of optimisation techniques to keep their cloud bills down.)

What is expensive? It cost $8 an hour to be online with the machine, which is why the kids stayed offline for as long as possible, only logging in once they were ready to submit their programs. Once they did, it cost them eight cents per second for computer processing time (Manes & Andrews 1993). All this was much cheaper than buying a computer but, since the average monthly salary in 1968 was $641, it wasn't cheap, and if a student accidentally plunged a program into an infinite loop, it was easy to rack up a bill for $50 before realising anything was amiss (Allen 2012).

Time-sharing was cheap enough to bring computers to those

who would have otherwise been excluded. It was not, however, cheap enough for the kids to spend all their time on the machine. For them, conversational interaction was available, but not all the time.

A lifeline

The Lakeside kids had the magical power of computing in sight but, due to the limited budget of the Lakeside Mothers' Club, were only allowed to touch it intermittently. In the middle of the academic year, however, their luck was about to change.

A new company, Computer Centre Corporation (Triple-C), was about to start selling access to a time-sharing computer. Before they switched on access for their customers, they wanted to test the machine, and who better to test it than the kids at Lakeside? They had, after all, already mastered BASIC and the GE-635. Triple-C made an arrangement with the school. They got free labour and the kids got what was only available to them in their wildest dreams: unlimited computer time.

The arrangement worked well while it lasted. The kids found bugs that nobody else would have looked for and Triple-C fixed them. At some point, however, Triple-C had to declare that testing was over because they desperately needed to provide access to fee-paying customers. The kids' access was revoked and in return for their efforts they received a paltry 20 per cent student discount should they want to log back into the machine again.

Cybercrime and punishment

Predictably, the kids were not ready to give up the dream. During testing, they found a chess game that pushed the computer to its limits. They then spent thousands of dollars' worth of computer time playing it. The kids, as easily as they found the chess game, also found the files that contained usernames, passwords and billing information.

Once they had the files, they looked up a test account that had unlimited and free access to the mainframe. They logged in with that. It must have seemed harmless to them, not least since nobody would ever have known about this and, besides, they were only doing what their mums had encouraged them to do.

Their cunning plan almost worked, but their meddling maths teacher noticed something was amiss. He called Triple-C to share his concern. Triple-C were nonchalant. What damage could a bunch of kids possibly do? They made a change to the operating system to make the computer impossible to break into. The kids broke into it in 30 minutes.

After school that day, they went over to Triple-C to show them how they did it. This might have all gone unnoticed, something the kids and the engineers at Triple-C kept to themselves, but soon the most senior executives got wind of what had happened. They were not happy.

Triple-C were publicly embarrassed. Their system was shown to be pathetically vulnerable. But privately, they were relieved. The kids had found serious flaws in their system's security.

The school, on the other hand, was humiliated. The privately educated kids of the most well-to-do parents in Seattle, aided and abetted by the Lakeside Mothers' Club, had committed, if not a crime, then at least some sort of misdemeanour. A line had been crossed. Justice had to be served.

The parents, teachers and executives from Triple-C got their heads together. They came up with a punishment that they felt fit the crime. The kids were dragged before the principal and what they thought were two FBI agents. The FBI agents were in fact managers from Triple-C. The Lakeside kids were then given an agonising punishment. They were not grounded. They were not fined. They were not suspended from school. They were instead forced to join the millions of withered souls who had known, and who would come to know, the anguish and humiliation of having the joy of computing ripped away by an arbitrary wielder of power such as the high priests of the mainframe or worse, the goddesses of computer access who reign supreme in all fledgling programmers' lives: their mothers. The kids were banned from the computer room and thus doomed to the tortuous prospect of having to *play outside*.

This punishment was so traumatic that one student later went on record to say that the ban was for a whole year (Manes & Andrews 1993). It was, in fact, only for the summer of '69.

Where to now?

In 1968, utility computing arrived at Lakeside. It would, by the 1970s, become the prominent model of computing and, as it turned out, it didn't require a nuclear-proof communications system for it to work. Instead, it worked on the telephone network and had its signals from the dumb terminals to the mainframes amplified by vacuum tubes.

As the 1960s turned into the 1970s, people could be forgiven for concluding that it would only be a matter of time before everyone had a dumb terminal in their home from which they could happily write programs. If that had happened, the cloud would have arrived 40 years earlier than it did. However, the tectonic plates of computing were shifting again and time-sharing computers, organised as a utility like the telephone, would suddenly seem naive, if not completely daft.

At the moment the Intergalactic Computer Network was bringing the internet to life and Bob Taylor was inventing the personal computer, the other Bob, Noyce, was about to do something that would change absolutely everything. In doing so, he helped to dump time-sharing straight into history's dustbin. You couldn't make up what happened next.

Chapter 9

The dramatic fall of utility computing

With the arrival of time-sharing, it looked as if computing would go exactly the same way as the telephone and electricity, because these two things had something in common with each other and, before 1971, with computing.

The telephone and electricity systems were expensive. Bell Labs spent inordinate amounts of money perfecting woodpecker-proof telegraph poles, wires and amplifiers that carried analogue signals further and further. The development of the phone system only made sense if the phone companies spread the costs of its constant renewal and maintenance over millions and later billions of users. This was Ted Vail's vision and legacy.

The same could be said of the electricity system, which was both the bottle (the grid) and the genie (electricity). It was expensive too but, unlike the phone system, electricity could, and in the beginning sometimes did, kill its users. This is why so much of the grid was about safety, especially at the point of consumption, where Zeus's thunderbolts made contact with their new human overlords (Schewe 2007). Like the phone system, the cost of maintaining and improving the grid also had to be spread over millions of users.

Computers were also remarkably expensive and that explains why utility computing in the form of time-sharing was so successful in the 1970s, briefly becoming the dominant form of computing. It's also why organising computers as a utility had caught the imaginations of the computer pioneers of the day. However, right at the start of time-sharing's rise to dominance, computers all of a sudden stopped being expensive and, unless a big one fell on somebody's head, they were unlikely to kill anyone.

In 1971, the logic for creating a computing utility collapsed. Time-sharing systems, those wonderful proto-clouds and the companies who built them, were about to come to a sticky end. Something very small but wholly fatal to their businesses was about to come crashing down on their heads.

Busicom

After he invented the microchip, Bob Noyce left the company he was working at, Fairchild Semiconductor, and started his own company with Gordon Moore, called Intel.

Intel was not in the computer business. It was in the chip business and specifically the memory chip business. It tended to avoid logic chips that did specific things because the market for them was limited (Ceruzzi 2012). Logic chips were also harder to manufacture, whereas computer memory, made up of row upon row of transistors, was comparatively easy.

Intel did, however, entertain an enquiry from Busicom, a Japanese manufacturer of calculators. Busicom wanted to create a wide range of new products, each with different applications, some of which at that moment were unknown. Busicom asked Intel to design a set of chips that would cover the entire range of mathematical functions that they needed for their new product range.

Rather than come up with specific chips to solve each mathematical function, Intel had the idea of designing a general-purpose chip whose instructions, or program, could be stored in memory chips, from where they could be loaded. In this setup, each mathematical function would be stored on a different memory chip and, when they were needed, loaded into the general-purpose chip and then executed.

In this way, Busicom would get all the mathematical functions they needed on memory chips and then get the general-purpose chip on which these stored programs would be executed. The chips could then be arranged into a circuit that was small enough to fit into a desktop calculator, or whatever new contraption Busicom had in mind. In the future, should Busicom need a new mathematical function for a new but unforeseen product, Intel could design and then produce for them a new memory chip with the program on it without having to redesign the general-purpose chip.

The programs on these memory chips were exactly like programs stored on punch cards. Some of them could calculate logarithms. Others might help solve differential equations. The difference was that the general-purpose chip loaded and then executed them easily. And the memory chips were, of course, much smaller than the reams and reams of punch cards that equivalent programs were usually stored on.

Because the instructions had to be loaded from memory, any machine with these general-purpose chips would be dramatically slower than other electronic devices. The designers at Intel knew, however, that the few milliseconds that elapsed between keystrokes on a desktop calculator were more than enough to load the instructions from the memory chip and into the general-purpose chip. NASA needed their guidance chips to work in real time but any chips whose operations involved humans thinking before their gigantic fingers stroked the keys could afford to be a lot slower. This was, without doubt, a beautiful response to Busicom's request.

A teeny-tiny computer

If programs stored in memory chips were exactly like programs stored on reams and reams of punch cards but almost infinitely smaller, and if the general-purpose chips executing those programs were exactly like mainframes but also vastly smaller, then hadn't Intel just invented a teeny-tiny computer? That's exactly what they had done – they just didn't know it. What Bob Noyce did know, however, was that if Intel owned the rights to the general-purpose chip, they could later sell it, along with new programs stored on memory chips, to other companies.

The ever-savvy Noyce brokered a deal. Intel would provide Busicom with all the chips they needed for a reduced price. In return, Busicom agreed to grant Intel the rights to market the chips to other customers, but only if they promised not to sell them to any of Busicom's direct competitors. In other words, Intel would not sell the chips to any companies in the calculator business.

Intel could now help all kinds of companies with their logical problems by storing the program on a memory chip and providing the all-purpose chip to execute it. These uses ranged from traffic control systems to pinball machines. The previous, tiny market for individual logic chips had been flipped on its head. It was now an infinitely large market. If a problem's solution could be stated as a computer program, then it could be loaded onto a memory chip by Intel and executed by its general-purpose chip. These general-purpose chips became known as **microprocessors**. This is how, unbeknown to Busicom and Intel, the most significant moment in the history of computing came to pass.

Microcomputers

In the early 1970s, there were two converging forces at work (Ceruzzi 2012). Users of time-sharing computers, like the kids at Lakeside, had experienced interactive computing and come to the same conclusion as their designers: time-sharing could form the basis of utility computing.

The other force was the relentless march of the team at Intel, who not only followed Moore's law, but saw it as a prophecy whose sacred duty was theirs to fulfil. This is how Intel had gone from the prototype microchip Noyce had made in 1959 to the general-purpose computer they had crammed onto a single chip in 1971.

These two forces needed to be brought together. But how? Well, once the microprocessor arrived, there really was no need to connect the dumb terminals of time-sharing systems to a mainframe. Instead, the keyboard could be connected to Intel's microprocessors and BASIC could be loaded onto a memory chip. This would give users totally unfettered, free and personal access to a computer that, since Intel's chips cost only a few hundred dollars, would be cheaper than renting a slot on a time-sharing computer. This was the dream not just of the Lakeside kids but the dream of everyone who had ever programmed a computer.

Fumbling the future

The most obvious candidates to capture this moment and thus write the history of the personal computer were IBM, one of its competitors or Intel itself. Unbelievably, none of them had the vision to do this because none of them saw the power of the microprocessor. Intel thought that their general-purpose chip would be embedded in products such as Busicom's calculators. It never seemed to occur to them that this tiny computer could be used as an *actual computer*. At the same time, IBM, whose revenue in 1970 was $7.5 billion, saw no threat from what looked to them like an inconsequential development. This lack of paranoia would cost them dearly.

The smashing together of these two titanic forces therefore fell to someone else; 25 years after Oppenheimer tested his bomb, the history of the world was once again going to turn on the axis of events in the deserts of New Mexico. Once again, it would not be turned back.

MITS

In 1970, Ed Roberts started a company called Micro Instrumentation and Telemetry Systems (MITS). The company sold model rocket kits and, in 1971, launched its first successful product, an electronic calculator. Roberts was one of thousands of hobbyists who, like the Lakeside kids, nearly all the boffins at Bell Labs and Noyce himself, loved tinkering. It was the hobbyists, and not the mighty IBM or the plucky Intel, who foresaw the two forces of computing coming together and Roberts who won the race to create the world's first **microcomputer**, the Altair.

Intel inside

Intel sold their microprocessors for $360 each. This price poked fun at IBM: their flagship, gigantic computer at the time was the System/360. Roberts was never going to pay that for them. He made a deal to buy the microprocessor for $75 each (Manes & Andrews 1993). This meant that MITS could sell the Altair for less than $400 in kit form or a bit more if the customer wanted it fully assembled.

Not surprisingly, Roberts's remarkable machine made the January 1975 cover of *Popular Electronics*. The headline read: 'World's First Minicomputer Kit to Rival Commercial Models' and off to the side of a picture of the machine it said 'Save Over $1,000'.

This was a mind-blowing and history-making headline. It was also wrong. The Altair wasn't $1,000 cheaper than its closest competitor – it was several thousand dollars cheaper. There was nothing like the Altair, nothing as small and absolutely nothing like it at that price.

Ed Roberts was within touching distance of bringing the two titanic forces in computing together. All he needed to do was to get a version of BASIC onto his machine that *Popular Electronics* already called 'magic'. This would give his customers a way to get busy with the Altair like the Lakeside kids had once got busy with the GE-635. Unfortunately, Roberts had no idea where that BASIC was going to come from.

Microcomputer software

In the 1950s, the pioneers who built and used computers started to realise that computer programming was a thing, that it was important and that it was worth investing time into tools that made it easier to do (Ceruzzi 2012).

To differentiate programs from the computer hardware they controlled, developer tools such as BASIC interpreters, application programs, such as the one that Lyons used to calculate tea blends, and operating system programs such as DTSS, which unburdened the programmer from managing the machine themselves, became known as 'software' (Ceruzzi 2012).

Roberts was crying into the dark for software. His cry was about to be answered. Since they graduated from Lakeside, the two ringleaders of the computer heist continued their obsession with computers and the software that controlled them. Their mums were still telling them what to do, which is why, when that edition of *Popular Electronics* hit the newsstands, they were studying for practical and sensible undergraduate degrees.

Their mums, however, couldn't stop them buying an Intel microprocessor, which they used as part of a system they built for measuring traffic data. It was called Traf-O-Data and was exactly the sort of application that Intel thought their programmable chips would be used for.

In order to write software for Intel's microprocessor, the Lakesiders programmed a time-sharing computer to simulate the whole chip. Using software to simulate hardware in this way meant they could write and test their Traf-O-Data programs on the mainframe. This made it a lot easier and more productive than interacting with the microprocessor directly.

Because of their simulator, the experience they had building Traf-O-Data software for Intel's microprocessor and their undeniable expertise with BASIC, when the Lakesiders saw the *Popular Electronics* headline in December, they were confident they could create a BASIC interpreter for the Altair.

The Lakesiders come of age

Neither of the Lakesiders wanted to speak to Ed Roberts, so they came to an arrangement. The youngest one, Bill, whose legs had swung under the chair in the computer room at Lakeside, would make the call. He put on the most grown-up voice he could muster. If they survived the phone call, the older one, Paul, would make the trip to Albuquerque to demonstrate their BASIC to Roberts. The call went well. Luckily for Bill and Paul, Ed Roberts didn't have enough memory cards for the machine. Would Paul mind coming over in a month's time?

The race was on, because they weren't the only programmers who thought they could write a BASIC for the Altair. Paul focused on altering the simulator so that it could simulate the microprocessor in the Altair, which was Intel's next version, the 8080. Bill got busy with the BASIC interpreter. None of this was easy but they had enough experience as programmers as well as working ridiculously long hours at dumb terminals, like they once did for Triple-C.

They hit the deadline. Paul made it all the way to Albuquerque to meet Roberts, where he demonstrated the BASIC, which was programmed into the Altair through shaking hands and, because it had no keyboard, its funny switches. The demo gods smiled on Paul that day – their BASIC worked.

This was how Paul Allen and William Gates, the one-time enfants terribles of Lakeside's computer room, collaborated with Ed Roberts to bring the two titanic forces of computing together. They had created a genuine, and cheap, microcomputer.

The Altair and Alto get married

The Altair was a direct descendant of ENIAC. ENIAC was created to help with firing tables and later computers like it were used to predict elections and, in England, calculate tea blends. These machines were about taking inputs and, through a series of calculations, creating outputs. They were not interactive.

Later, through the magic of time-sharing, these machines were made not fully interactive but at least conversational. The Altair didn't have a user interface or even a keyboard. Paul Allen inputted the BASIC interpreter using switches. It was nonetheless designed in the tradition of time-sharing computers. Paul and Bill were, after all, chasing the sort of experience they once had with General Electric's 635.

Xerox's Alto, on the other hand, was a direct descendant of radar. It was designed with interaction in mind and born out of Licklider's 'Man-Computer Symbiosis'. This is why it had a screen, a mouse and a graphical user interface.

The Alto foreshadowed the future of how humans and computers would interact. When it was finally commercialised as the Star, it was networked to other workstations using Ethernet, used a windows-based graphical user interface, had folders, icons, a mouse, file and print servers, and shipped with email as standard.

The Altair, on the other hand, foreshadowed the future of the computer's physical architecture. It wasn't a powerful or even useful machine but it demonstrated the possibilities and potential of cheap computing. Those possibilities were all about using a single microprocessor inside of a computer that had an operating system and other software pre-installed on it.

These two futures came together in the next generation of microcomputers whose hardware design evolved from the Altair and whose user interface and features evolved from the Alto.

The most famous, but certainly not the only example of the marriage of these two ideas, was the Apple II. The Apple II arrived four years before Xerox got the Star out of the door and contained almost all its features.

A BUNCH of misjudgements

In 1971, the key players in computing were IBM and the 'BUNCH' companies – Burroughs, UNIVAC, NCR, Control Data and Honeywell. IBM's market share was about 70 per cent.

The BUNCH companies, RCA and General Electric (Edison's company) shared the other 30 per cent.

Collectively known as IBM and the seven dwarfs, along with Xerox, they were wiped out or at least seriously wounded by bullets that were fired in the first half of the 1970s. The first bullet was the microprocessor. The second bullet was the Altair.

The world's first microcomputer managed to bring the 'magic' of time-sharing and microprocessors together.

The arrival of the microprocessor was the exact moment in history when software became much more important than the hardware it controlled. Real utility from that point on would come not from the machines themselves but what could be done with them. This is what Busicom and Intel stumbled upon and Ed Roberts, Bill Gates and Paul Allen took advantage of. Why did IBM and the BUNCH companies fail to see this?

They made two mistakes. The first was that they didn't understand Moore's law or, if they did, they didn't believe it. If they had, they would have known that the chip at the centre of Ed Roberts's woefully underpowered Altair could be replaced every 18 months with one that was twice as powerful. If that happened, it would only be a matter of time before their monolithic machines would be the size of a foot locker that could fit under a desk. The future was not going to be in big computers or even small ones. The future was going to be in microcomputers.

The second mistake was that they didn't understand the market. IBM, the dwarfs and Intel could not imagine a world where normal, everyday people would want to program or interact with a personal computer. After their experience in 1968, Bill Gates and Paul Allen laboured under no such illusion. With the right software, they envisioned computers would help busy housewives store recipes, stockbrokers access prices and they would keep bored children entertained with games.

As the cost of chips came down, time-sharing vanished and humanity's computing needs were met with personal computers. The personal computers available in the stores in the mid-1980s looked like the machine Bob Taylor had envisioned in 1972.

Where to now?

By the end of the 1970s, ARPA had completed the internet and showed that it worked.

At the same time, thanks to the work of PARC, the architecture for modern computers, including mice, keyboards, graphical user interfaces and word-processing software was more or less complete.

By that time, because of Ed Roberts's insights, it was obvious that computers didn't have to be big or expensive but could instead be built around just one microprocessor. Since Moore's law never stopped, each generation of personal computers was destined to be more powerful than the last.

This meant that when 1985 rolled around, personal computers were cheap enough for everyday people and powerful enough to be used by kids to play games, programmers to write new applications and mums and dads to balance the family's budgets. They were used by businesses all around the world which, without any help from Xerox and in the spirit of J Lyons & Co, had created their own electronic offices.

By 1985, almost all the pieces for wide-scale, distributed and networked computing were in place. It was only a matter of time before these pieces would come together, as they had once come together for the electricity grid, and in doing so give to the human race a worldwide computer system that was hitherto unimaginable.

Humanity, at this point, had long since stopped raiding Mount Olympus. The gods had nothing left to steal. The bringing together of all these components therefore fell to a frustrated scientist who, despite his organisation's ability to accelerate particles to the speed of light, could not look up phone numbers any quicker than he could find a new pair of shoes in a Sears and Roebuck catalogue. This scientist came up with a funny idea for a computer program. He called it the World Wide Web.

End of Part 3

It's easy to understand how punch cards can be used to turn switches on and off to either make a fairground organ play music or to complete circuits to power microchips. It's much harder, however, to fathom how program code, even written in a high-level language like BASIC, can be translated into an elaborate dance of billions of transistors flipping on and off at the speed of light.

This is why I believe that 1971 is the moment when the mechanics of computers and their users became alienated from each other and the word 'magic' started to be used for something that, as we've seen so far in this book, isn't really magical at all. The computer is composed of simple concepts and its 'intelligence' emerges from the speed at which it carries out equally simple operations. It's nothing to be feared. Yet, once thousands of transistors later gave way to millions hidden inside a quantum prison of silicon, they became impossible to fathom for everyday people. Mind you, this didn't matter to programmers like Paul Allen and Bill Gates. They cared little for how the computer worked and much more about the tunes they could make it dance to.

At the moment Moore's law started to noticeably kick in, just after 1971, software started to become more advanced, and from that point on could make millions and later billions of transistors do the most remarkable things. Once complex software met an ever-growing (and impossible to imagine) number of transistors, the conversational nature of user interaction with computers, which started with time-sharing, took on incredible forms and created the illusion that users were not only talking with machines and they were talking back but that maybe, just maybe, they were thinking, too.

These machines arose from an intense moment of procreation between ideas. Mainframes were brought together with operating systems such as DTSS to create time-sharing. Time-sharing itself was brought together with the microprocessor to create microcomputers. And the marriage of network technologies and computers gave us the internet. Later, the features of the Alto were brought together with microcomputers to create machines such as the Apple II.

By the end of the 1980s, there were two more scintillating forces that needed to be brought together. The first was the personal computer. By 1989, thanks to Moore's law and advances in software, personal computers were practical, affordable and powerful. The second force was the continued development of the internet and the rolling out of, and integration with, fast networking technologies such as fibre-optic cables.

The bringing together of these forces would be the final marriage between ideas whose offspring would finally bring Licklider's vision of a network of computers working in symbiosis with people to life. The bringing together of these forces fell to Tim Berners-Lee.

The web

Tim Berners-Lee worked at CERN, which stands for Conseil Européen pour la Recherche Nucléaire (the European Organisation for Nuclear Research). In 1989, there were 10,000 people working at CERN, each of whom had a paper copy of a gigantic telephone directory. Those born in the age of electricity cannot imagine a world without the grid; those born after the 1970s cannot imagine a world without digital communications; and those born after 1990 cannot imagine the type of mad world I was born into that actually contained paper telephone books. Nevertheless, those books and that mad world existed, and at CERN they existed in an organisation that was full of computers that were *on the same network*.

Berners-Lee had an idea for a system that allowed users to share files, regardless of their format, in a 'World Wide Web' of information.

The files could contain data, images, text or even movies and music. Berners-Lee's web allowed users to share files from their computer, whether they were the results of a new experiment, a picture of a cat or the telephone directory, with anyone else on the web.

There were two sides to the web equation. On one side was the web server. This was a computer whose software 'served' web pages. Web pages could be made up of text and, if the person who wrote the page wanted them, links to files. The people who wrote CERN's telephone directory, for example, could write a page with everyone's phone numbers in it and 'serve' that as a web page. In fact, to help prove the concept, Berners-Lee convinced CERN's telephone administrators to do just that.

On the other side of the equation was a web browser. This was a simple bit of software that 'spoke' to the web server, asked for pages and then presented them to the user, who could be on the floor upstairs, on a different campus or even on the other side of the world. Using what was called a 'unique resource identifier', the browser called up a page such as the telephone directory and displayed it in a way that made it easy to read and, by clicking on links, easy to 'browse'. The resource identifier had to be unique because every piece of information was unique, too.

Was Berners-Lee's web any different from the telegraph itself? In some ways, it was not. There were two machines, computers in the case of the web or transmitters and receivers in the case of the telegraph, and messages were converted into unique pulses of electricity before being sent along a wire only to be converted back on the other end.

In some ways, however, it was very different. Once the signal was moving across the internet, it was doing so in packets that were routed across a decentralised network by other computers. A nuclear strike would have stopped messages moving over the telegraph lines but would not have stopped Berners-Lee browsing CERN's phone directory or his mum sharing pictures of the cat.

Like Gray's utterly bonkers needle in an electrified solution,

Bardeen and Brattain's first transistor with its chunk of germanium with wires stuck into it and the doomed ENIAC and its squadron of maintenance engineers, the first browsers, including Tim Berners-Lee's, were terrible.

That changed in 1992. A student from the National Centre for Supercomputing Applications (NCSA) called Marc Andreessen built the first really usable web browser. Called Mosaic, it was ready for its first release in February 1993.

Just over a year later, Andreessen quit NCSA to set up a company called Netscape. Sixteen months after that, on 9 August 1995, Netscape became a publicly listed company. After the first day's trading, despite making no profit at all, it was valued at $2.9 billion. This gave rise to the term 'Netscape moment', which has been used to signal the birth of a new industry. This moment, however, did not only signal the birth of a new industry. With hindsight, it's clear that it signalled a new epoch in human history.

Themes redux

In Part 3 we really did learn a lot about computers. We now know why some programmers call their craft coding. We know software got its name because it's not hardware. We learned the origins of the word modem. Through the work of Bob Taylor and Ed Roberts we also saw the computer shift its shape again. Calculating machines such as ENIAC and real-time systems such as radar merged and reappeared as the personal and networked computer.

We also got a much better understanding of humanistic management because we learned about Peter Drucker's symphony model. Bob Taylor, like Oppenheimer before him, took on the role of orchestrator. For innovation's sake, Taylor fought bureaucracy and compartmentalisation.

Finally, a new theme emerged:

> The impact of new technologies is almost always overestimated in the short term and underestimated in the long term.

This is Amara's law (named after scientist and futurist Roy Amara). It explains why the excitement around the transistor's invention faded as it had once done for the vacuum tube. The moment between the excitement fading and the real power of a technology explains the rather silly dismissal of the telephone by Western Union. It explains why nobody really predicted the impact the electricity grid, for example after its initial inception, would have on society. It explains why Intel and IBM could not see what Ed Roberts clearly could. It explains why the Nobel Prize-winning economist Paul Krugman wrote in 1998 that 'by 2005 or so, it will become clear that the internet's impact on the economy has been no greater than the fax machine's' (Krugman 2023). Nobody could blame him for saying that. The web was pants and would stay pants for years. However, after the hype and inevitable disappointment, technologies have a way of really changing things and we humans tend to underestimate just how much. I think Amara's law explains why many people are making the same mistakes with AI right now, just as we once did with the web. That's a mistake I personally don't want to repeat.

Where to now?

Once Netscape arrived, it was inevitable that entrepreneurs would try to commercialise the internet using the World Wide Web. That means we're right back where we started. The year is 1994. The place is New York. Jeff Bezos is about to set in motion a chain reaction that began with ecommerce, ended with artificial intelligence and in between brought time-sharing back to life, only this time round, in the hands of canny marketeers, it would be called the cloud.

Part 4
CLOUDIFICATION

Chapter 10

The magical computer store

David Shaw was Jeff Bezos's work soulmate. Deciding to leave D E Shaw & Co was therefore not straightforward, even for a man who thought (and probably dreamed) in algorithms.

D E Shaw & Co

David Shaw graduated from Stanford with a PhD in computer science. He hopped to New York's Columbia University before skipping to Wall Street, where he hoped to make one last jump to entrepreneurship and his fortune. That jump occurred when he started D E Shaw & Co on top of a communist bookstore in Manhattan (Stone 2014).

D E Shaw & Co was different to other Wall Street firms because David Shaw was different. He wasn't a finance type but a left-leaning computer science professor who had used the ARPANET in his university days and shared Licklider's passion for a network that bound the world's computers together. To the outside world, D E Shaw & Co may therefore have looked like an investment firm but to Shaw it was a technology company. Investment just happened to be the first arena that the brainiacs he hired to program his computers would play in (Stone 2014).

This is why it wasn't totally left field, when the internet finally reared its head from the obscurity of academia and the Department of Defense, that D E Shaw & Co wanted to try to make some money with it. Figuring out that money making fell to a pasty-faced, hard-working, intelligent (but not intelligent enough to study theoretical physics) and, although not extroverted, the least introverted man at D E Shaw & Co. That man was called Jeff.

It didn't take long for ideas to start flying around. One was a web-based email service. Another was a system that traded stocks and bonds for online users. A third was an application that arbitrated deals between every type of buyer with every type of seller, albeit with an egalitarian twist: users would leave reviews, and so, as well as being profitable, this 'everything store' would be the focal point for a community of real users (Stone 2014).

Beginnings

The everything store was a clever idea. Such a clever idea, in fact, that in 1994, Jeff Bezos left D E Shaw & Co and took it with him. He toyed with a few ideas for names. Cadabra from 'Abracadabra' was on the list but sounded too much like Cadaver (Bezos 2021). Gordon Moore and Robert Noyce had had a similar problem with their first name for Intel, Moore-Noyce. It sounded like More Noise, the last thing any semiconductor manufacturer wants to be associated with (Tedlow 2006). MakeItSo.com was a nod to his favourite TV show, *Star Trek: The Next Generation*. As a child he acted out scenes from the original *Star Trek* with his friends. They played Kirk and Spock. He played the computer (Randolph 2021).

Relentless.com was one of his favourites but it had sinister overtones, although not sinister enough to stop him registering the domain and redirecting it towards Amazon.com, something it still does today.

Bezos finally settled on Amazon after reading the dictionary. The Amazon was not only the world's biggest river; it was by far the biggest river. It blew every other river away (Stone 2014). It also conveniently began with an A.

Bezos's hacks

In 1994, there were no search engines like Google or even Ask Jeeves. Instead, users browsed web pages that aggregated useful links to sites and resources, such as Jerry and David's Guide to the World Wide Web, which would later become Yahoo.com (O'Regan 2008). Some of these internet guides arranged their listings alphabetically so Amazon.com landed close to the top of the rankings, which funnelled traffic back to the website.

This was not Bezos's only hack. Amazon's first warehouse was not a brand, spanking new building but the garage at his house. Amazon didn't bother with a fleet of branded trucks either, which a company called Pets.com did, preferring to use existing companies and the Post Office for shipping. In a further twist of irony, the communist bookstore being the first, Amazon's staff meetings and coffee breaks were held in the cafe at the local Barnes & Noble.

Bezos also knew the law. In 1992, the Supreme Court ruled that sales tax was not applicable in states where companies had no physical presence. By incorporating in states with small populations, mail order companies got away with paying just a sliver of sales tax for the small amount of goods they happened to sell in them. This is why Bezos headquartered Amazon in Seattle, Washington.

Amazon's low costs, thanks to the Supreme Court's decision and the savvy use of existing companies' infrastructure, including Barnes & Noble's coffee shop, were translated by Bezos into significant discounts, which was how he got customers like me out of the bookstore and onto his website. When those low costs met the staggering reach of Andreessen's browser and Jerry and David's guide, the foundations for a new type of business were laid.

From the bookstore to the everything store

David Shaw dictated where the creative energies of his introverts went. So too did Jeff Bezos. He hadn't given up on the idea of building the everything store but had to choose a place to start, a place to direct his energies (Stone 2014). Books had distinct advantages. The first was that they came from just two distributors. Acting as an intermediary between a couple of distributors and millions of users was going to be easier than dealing directly with thousands of publishers. The second was that books bought online were exactly the same as those bought on the high street or at the airport. Customers knew what they were getting and although at first they may not have trusted Amazon, they certainly trusted their favourite publishers. Finally, the web browser pointed at Amazon.com allowed customers, from the comfort of their own homes, to browse more popular and specialised books than would fit in a Barnes & Noble the size of Yankee Stadium (Gordon 2016; Bezos 2021).

Once it had the trust of its customers, and with Barnes & Noble's new (but inferior) website nipping at its heels, Amazon sidestepped into selling music, DVDs, electronic goods, software, games and toys. When that happened, it began its transformation into the everything store, exactly as David Shaw and Jeff Bezos had envisioned back in New York.

The return of the priesthood

Success pushed Amazon out of Bezos's garage and into progressively bigger offices. One office move saw Bert and Ernie, Amazon's computers, driven across town in the back of an Acura Integra (Stone 2014). In the beginning, Bert and Ernie were powerful enough to serve all of Amazon's users. By the turn of the century, however, Amazon needed lots of computers, data storage and networking equipment. In other words, it needed lots of IT infrastructure. A simple website can be hosted on a couple of computers but a web-scale application such as Amazon.com, with millions of users, cannot.

As well as needing all this IT infrastructure, those building web-scale applications had to learn how to manage it all because, in 2000, if the computers or the software running on them failed, then their businesses would start to fail too. So Bert and Ernie were eventually replaced by powerful computers and supporting infrastructure. Like mainframes before them, these new computers had to be stored in climate-controlled environments complete with fire suppression; had to have failover mechanisms such as backup electricity generators; and they had to be maintained because, although they didn't fail as often as the vacuum tubes in ENIAC did, they nevertheless did fail.

As companies like Amazon started to scale, their computer rooms started to resemble those of the 1960s and, not surprisingly, a priesthood and accompanying power struggle emerged around IT infrastructure as it had once emerged around mainframes.

Fettered and limited access

I experienced this madness first hand when I found myself building a web-scale application for an airline. At first, airlines used the web as an extra channel to communicate with their customers, like the White House did with pictures of Socks, the First Cat. It didn't take the airlines long, however, to come to the conclusion that if Amazon could sell books online, then they could sell plane tickets online, too.

One of the teams building the airline's web-scale application had got into a muddle and my boss thought I could unmuddle it. This was how I was introduced to web-scale application development and, after only reading about them in books, the high priests of mainframes who had reemerged as the high priests of IT infrastructure.

Our job was to write the software that allowed users to check schedules and book flights through a web browser. If the airline got this right, from that moment on customers' needs would be served without them having to walk into a real, high street travel agent.

This was an exciting time in the history of computing. We should have been sprinting. Instead, we plodded into the future with that booking system. Why? We didn't have direct access to the web servers where the system would eventually run. The only way to get our hands on those types of machines was by booking a slot. We did that by filling in a form and sticking it into an operator's in-tray. A few days later, we were allocated a slot on the web server. The slot was always weeks in advance.

In practice, this meant that if we built a new feature, we would have to wait weeks to find out what would happen if 100,000 users tried it at the same time. If the test failed – and, like punch cards, they always failed the first time – we'd fix the code, get back in the queue and run the tests again weeks later.

When you had direct access to the web servers, like Amazon's programmers once had access to Bert and Ernie, the time it took to test newly completed software was measured in minutes. Because of limited access to the computers, the time it took to build and test features of the web-scale applications we were building at the airline was measured in weeks and sometimes months.

After having free and unfettered access to computers for most of my life, I was all of a sudden at the mercy of the high priests of IT infrastructure who dictated the pace of our work in exactly the same way that they dictated it at Amazon. The problems of mainframes had somehow been brought back to life and the ghosts of a long-dead generation weighed like a nightmare on the minds of those still programming.

Two forces

At the turn of the century, there were two forces swirling around Amazon. The first was its laser-like and relentless focus on customer value. This meant, in practical terms, that it was frugal. It counted dimes like all high-volume, low-margin businesses do. The second was its desperate desire to become a genuine technology company so it could break away from its stock price and its humble roots in retail (Stone 2014).

It was by dealing with its IT infrastructure, which was expensive, dictated the pace of innovation and heaped misery on an already miserable workforce, that Amazon would on the one hand optimise its low-margin business and on the other start the journey to becoming a genuine tech company. At the turn of the century, without much of a clue as to what it was doing, Amazon found itself teetering on the brink of performing a miracle.

Infrastructure as a service

Amazon didn't have a master plan – not at first, anyway. Instead, it made simple moves to optimise the IT infrastructure bottleneck. The first move was to stop the sort of low-level and time-consuming coordination that plagued my team at the airline. It introduced a 'coarse grain' interface between the priesthood who provisioned IT infrastructure and those who built the software that would eventually run on it (Bezos 2021).

From that moment on, software developers at Amazon would send requests to the infrastructure teams, who would do their best to get whatever they needed back to them. In a reversal of power, the software developers at Amazon became the customers of the IT infrastructure teams whose job it was to facilitate their creativity.

It was through this service-oriented and developer-centric approach, something that was ingrained in its DNA, that Amazon started to optimise the IT infrastructure bottleneck for creativity (Bezos 2021). This way of organising infrastructure teams became known as infrastructure as a service (IaaS).

Standardisation

Amazon's second move happened in 2003. Infrastructure provisioning would, from that point forward, be standardised and fully automated. If it got this right, then the system used for provisioning IT infrastructure could be connected to a web application. If that worked, its software developers could order IT infrastructure through a browser just like its own customers ordered books on Amazon.com. Bezos summarised this new approach in a memo that contained just a handful of bullets. The second to last bullet said that anybody not following the previous bullets would be taken outside and shot. Just kidding. But it did say they would be fired. It was not a joke.

Throughout this gestation period, the team at Amazon became obsessed with an idea. They asked themselves, what applications might a student working in their dorm rooms and who had easy and instant access to the same infrastructure as the world's biggest ecommerce company come up with?

If Amazon kept improving and simplifying its system of infrastructure provisioning so that it really could be used by a student, it would eventually invent a magical computer store. Its own software developers would be able to create and test their ideas experimentally, tapping out web-scale applications as easily as I'd once tapped out BASIC on my Commodore 64. Amazon would have in its possession a system of innovation that its competitors could not envision let alone build.

Undifferentiated heavy lifting

As the magical computer store materialised in the minds of those at Amazon and as bits in the depths of its servers, Jeff Bezos and his assistant, Colin Bryar, attended the O'Reilly Emerging Technology conference. The founder of Flickr, a web-scale photo-sharing application, was on a panel. He told the audience that about 50 per cent of his team's time was spent scaling their IT infrastructure so that the company could keep up with its rapidly growing user base.

The team at Flickr, like my team at the airline, were in other words constantly scaling their computers, databases and adding web servers. None of this enhanced the application they were building. This meant that instead of grabbing land on the digital frontier, companies like Flickr spent half their time faffing around with their computers.

Bezos and Bryar realised that Amazon could have taken care of Flickr's infrastructure for them. Or, to use a term that emerged within Amazon, they realised that they could have done Flickr's 'undifferentiated heavy lifting' for them (Bryar & Carr 2021). It dawned on Amazon, embryonically at first and then epiphanically, that its magical computer store was not just a way to optimise its own low-margin business. It was a way to optimise *all* the world's businesses. If Amazon could do other businesses' undifferentiated heavy lifting for them, then they would be free to experimentally develop their web applications in the same way that Amazon's programmers developed theirs (Bezos 2021).

Amazon Web Services

After a relentless period of development, on 14 March 2006 Amazon began opening its magical computer store to the public. They started by letting customers rent storage from them. Not long after that, customers could rent computers, too. This meant that a student who wanted to create a web application to share photographs, like Flickr had once done, could do so without worrying about the upfront cost of data storage, the ongoing monthly costs of maintenance and the annoyance of having to replace their hard drives and machines when they eventually failed – as fail they all eventually do.

For the same reasons, Amazon's magical computer store was a great tool for start-ups and scale-ups, who now could access the IT infrastructure of the world's leading ecommerce store and use it for whatever applications their founders dreamed up. Amazon called its computer store Amazon Web Services (AWS).

In an interview in 2009, in an attempt to explain AWS in terms that were easy to understand, because lots of people still didn't understand it, Jeff Bezos said that AWS was analogous to the electricity grid. In the same way that factories had once dumped their power plants, modern businesses could now dump their computers in favour of computing infrastructure that was supplied on demand and, like electricity, as a service (Rose 2009).

Bezos had plucked the flower of technology out of the nettle of Amazon's own dilapidated IT. The old man had done it. He was 45.

Amazon's DNA

A book publisher criticised Jeff Bezos, telling him that his job was to sell books and not, through user-generated feedback, trash them (Kirby & Stewart 2007). Bezos begged to differ. His job was to help Amazon's customers make purchasing decisions. Bezos's ability to see things from a unique perspective set him apart. Whereas other bookstores thought their job was to sell books, he thought it was Amazon's job to assist with purchasing decisions.

Whereas other businesses hurried to consolidate their gains, Bezos realised that the web was an untapped frontier and gains made in a patient but relentless march would be his to exploit later. Besides, his customers paid him before he delivered their books and this cash flow helped Amazon conquer the digital frontier like Richard Sears and Alvah Roebuck had once conquered the actual frontier.

Whereas other retailers thought technology was an aid to doing business, Bezos's retail business was an aid to doing technology. His intuition to see things differently was honed at D E Shaw & Co but rooted in ARPA. Jeff's hero, Alan Kay, was a veteran of both ARPA and PARC. Kay said that 'point of view is worth 80 IQ points' (Stone 2014).

There was another, more direct influence on Bezos's thinking, however. That came from his maternal grandfather. His summers were spent on Lawrence Gise's ranch and they included the sort of hard, self-starting work you find on all ranches. However, they also included long conversations about technology and science. Gise was a well-qualified teacher, previously having worked at the Atomic Energy Commission, where he helped develop the hydrogen bomb (Bezos 2021). Between 1958 and 1961, just before Licklider arrived, Gise also worked at ARPA on ballistic missile and space technology (*Journal Staff Report* 1995). This goes some way to explaining the origin of Bezos's own obsession with technology and space travel (O'Mara 2020).

The direct influence of Shaw and Gise, and the indirect influence of Kay, is how I think the rebellious spirit of Licklider and ARPA found its way into Amazon's DNA. When that DNA came back out as AWS, it was possible to enter into human–computer symbiosis not with a single personal computer but with a whole 'cloud' full of them.

Where to now?

Amazon was almost a real technology company and the internet was almost commercialised, which, as inevitable as that seems now, was anything but in the mid-Noughties. What Amazon and the web really needed was a rising star that not only took advantage of Andreessen's browser and Jerry and David's guide but of AWS, too. What they got instead was a mail order video store whose business model was so incomprehensibly daft that it made the idea of selling books online look almost as clever as Bob Noyce's layer of silicon oxide.

Chapter 11

Saturday night at the movies

Many internet businesses failed not because of poor technology but because they had nonsensical business models, poor leadership and management, awful software development practices or wasted all their money on building fleets of trucks and networks of warehouses (which is what Pets.com did) or built a website that was so lavish that the pages wouldn't download on time (which is what Boo.com did) (O'Regan 2008).

In 1997, the pantheon of nonsensical business models seemed destined to grow by one.

Two men, one of them a seasoned entrepreneur, the other not quite so seasoned but who happened to be a distant relative of Sigmund Freud, decided to build an online mail order video store. His wife told him it would never work.

The conceptual miracle

Video stores had two serious deficiencies baked into their business models. The first was late fees. The second was poor access to new releases. The two founders of the online video store hated these deficiencies. My brother Brendan and I hated them, too.

Throughout the 1980s and the early 1990s, when we still used an old rotary dial telephone with Edison's carbon button and the top of our street was blighted by a three-storey building that held within it a monolithic mechanical contraption whose job it was to route telephone calls, Brendan and I used to get to the video store before it opened so we could get our hands on the new releases. This did not work if someone else had already booked the film or if another kid got there first.

As late as 1997, when I was already 21, we had to wait three or four weeks before we could rent *Titanic*, *Tomorrow Never Dies* or *Men in Black*. In those intervening weeks, we had to avoid hearing about whatever film we were tracking down from our mates who had already seen it. Because of these deficiencies, managing dissatisfaction was the bane of all video store workers' lives. Ray, who ran ours, did this by recommending every alternative as 'excellent'. You'd go in for *Mad Max Beyond Thunderdome* and come out with *Spies Like Us* because Ray had told us 'Yes, it's excellent, excellent', always saying excellent twice. Ray's advice threw petrol on the bonfire of our dissatisfaction. Brendan and I hated waiting for new releases, hated paying late fees for a movie we had only watched once and hated Ray's awful recommendations.

One of the founders hated all this so much that the idea for an online store came to him after he found *Apollo 13* in the cupboard and then had to pay a $40 late fee (Shih et al 2009). Like millions of others, we all dreamed of personal and unfettered access to movies.

The gadget

The rise of the patchwork of bricks-and-mortar video stores, the later ubiquitousness of Blockbuster, and therefore our problems getting hold of a copy of *Pump Up the Volume*, can all be traced back to Bell Labs when, in 1953, they received a visit from Akio Morita and Masaru Ibuka.

Morita, whose destiny was to become the 16th Morita to manage the family *sake* business, and Ibuka, who opened an electronics shop in 1946, were the co-founders of Tokyo Tsushin Kogyo, the Tokyo Telecommunications Engineering Corporation (Morita 1986). The problem for Morita's legacy was that he was not into *sake*. He was into gadgets (Miller 2022).

In 1948, he read about the transistor, saying it was miraculous. He wasn't wrong. Bell agreed to license it to Morita and Ibuka but warned them that, apart from the hearing aid, there were no other practical products that could be manufactured from transistors.

Morita thought the scientists at Bell Labs were wrong. Locked up in a noisy room in Taft's Hotel close to New York's Times Square, he and Ibuka intuited that transistors would be used in products that were at that time unimaginable, even to them. The transistor was, in other words, a solution looking for a problem (Morita 1986).

On the same trip, after seeing first hand how much disposable income everyday Americans had, which stood in stark contrast to their own war-torn and poverty-stricken country, Morita and Ibuka realised they had found the perfect country in which to sell the electronic products they planned to invent.

These two insights came together in their minds. The Tokyo Telecommunications Engineering Corporation would develop ingenious consumer goods using America's semiconductors and then sell them back to its spendthrift citizens.

Things started to fall into place for Morita and Ibuka. All they needed, because at the time Americans thought that everything coming out of Japan was junk, was a new name. In 1958, they combined the Latin for sound, *sonus*, with the American nickname Sonny, and got to Sony. The *sake* industry was going to have to get by without Morita. He was about to take advantage of the ever-shrinking cost of semiconductors and the ever-growing spending power of American citizens.

Legal battle

In 1975, with an advert that boldly said 'Now you don't have to miss *Kojak* because you're watching *Columbo* (or vice versa)', Sony brought to market something much more interesting than a hearing aid (Lardner 1987). They released the first commercially available video cassette recorder (VCR). Universal and Disney immediately began legal proceedings to ban its sale (Payne 2021; Greenberg 2008).

In 1983, Universal and Disney eventually lost. But it didn't matter. Nobody had bothered waiting for the Supreme Court's decision. Almost immediately after the VCR was invented, the adult entertainment industry worked out how to produce and distribute movies on video cassettes (Barrs 2012). Soon after that, benefiting from the trail the pornographers had blazed, the mainstream followed in their wake. This included Disney, which, despite being the plaintiff, did not wait for its own case against Sony to wind its way through the intestines of the legal system.

The next idea

With all legal avenues closed, having failed to kill the VCR, the studios had a chance to kill the video store before it was born and thus take the home movie market for themselves. All they had to do was sell movies directly to the public and in doing so they would cut out the middleman and keep all the profits to themselves. It was an open goal. It was impossible to miss. They missed.

Convinced that the price of *owning* a movie had to be higher than the price of renting a seat in the cinema, the film studios decided to sell their video cassettes for about $65 (Payne 2021). That's $250 in 2025 money.

The studios thought that this was a fair price to be the proud owner of a movie that could be watched in the comfort of a home. After all, NBC's *Saturday Night at the Movies* and the arrival of cable television had already domesticated movie consumption, sowing the seeds in one decade that would yield, to paraphrase a newspaper headline from the 1990s, a bumper crop of couch potatoes in the next (Herbert 2014). The studios were right about home consumption. But they were wrong about the price. Spendthriftiness may have been consigned to the previous century but consumer stupidity had not. Americans didn't want to be ripped off (Herbert 2014). They refused to pay and the video store industry was born.

Unintended consequences

It's not an understatement to say that the film industry was terrified. Steven Spielberg said that videos would take the magic out of the movies. Jack Valenti, the president of the Motion Picture Association of America, went even further, testifying to Congress in 1982 that the VCR is to the American movie-loving public what 'the Boston strangler is to the woman home alone' (Payne 2021).

They were both wrong. In 1988, Disney released *Cinderella* for $29.95. It sold six million copies and generated $180 million in revenue. *Fantasia*, a movie made possible all those years ago by David Packard and Bill Hewlett's audio oscillators, was re-released in 1991 on video cassette. In the United States alone, it sold 14.3 million copies, generating half a billion dollars of revenue in only 50 days; 263 million of those dollars were pure profit (Brew 2019). A few years later, in 1995, they made another half a billion dollars by selling 32 million copies of *The Lion King* (Payne 2021).

The predicted death of the movie industry was not greatly exaggerated; it was completely wrong. Sales of video cassettes were an untapped source of revenue that all the mainstream studios profited from. But they were not alone. Independent studios benefited, too. The increased choice and reach of video stores meant that smaller movies that wouldn't have been commercially viable a decade earlier were now getting made. *Reservoir Dogs* (1992) is the most famous example of this. Not only did it make millions at the cinema before appearing in the video store but it was written in one, when writer and director Quentin Tarantino worked in Video Archives on Manhattan Beach in California.

Economically, as early as 1987, the hodgepodge of video stores, some run as mom-and-pop stores, others as add-ons to gas stations and convenience stores, were contributing $3 billion to the US economy. When Blockbuster was sold to Viacom in 1994, it went for a mind-bending $8.4 billion.

Psychologically, the video store became a central place to

share ideas and meet, becoming a focal point for a community of consumers. Video stores habitualised the browsing of films, which people did alongside picking up their groceries or searching for books. Video stores amplified but didn't change existing trends for the domestication of movie consumption, the desire for a broad array of media options and shopping for tangible goods such as books and records, which merged into the retail habits of millions of people around the world (Herbert 2014). The relentless marketing of the studios also shifted consumer expectations. It created a sense of entitlement, which led to queues outside the stores and frustration when customers like me and my brother could not get their hands on the latest releases.

The immutable model

If the movie business was booming, what was so daft about starting a mail order, online video store? The problem for our two entrepreneurs was that there was money in movies but not in video stores. The studios were the big winners. The stores and their customers were the big losers. The price of a new cassette was so high that the video store owners had to rent out a new release every day for a month before breaking even on its cost. Hardly any store could afford to buy more than one copy. Demand for new releases therefore outstripped supply, hence the Saturday morning queues.

Because of the time the movie spent in transit, a mail order video store would be able to rent out a new release about four or five times a month. Well before a mail order store broke even on the cassette, the new release would no longer be a new release. The cost of postage eroded profit further while the damage to or loss of the cassette, once it was in transit or on the coffee tables of customers, was not only real but probable. It was a hopeless idea.

The transformability of information

Then, from out of the blue, our two video store entrepreneurs met with an extraordinary stroke of good fortune. The ability of information to shift its shape, to transform from any one form to another, explains how a movie can be projected onto a cinema screen or beamed as a television signal where it's later reconstructed through line-by-line blasts of electrons from a cathode ray tube. The transformability of information explains how the same information, on a video cassette, a totally different physical form, made its way into homes all around the world in the 1980s and 1990s. It explains the invention of the DVD in 1996.

Digital video discs (DVDs) were a serious upgrade for how information could be delivered into people's houses. What was previously available on cassettes was now going to be burned on a 12 cm disk. The studios, more than 20 years after they first tried, had another chance to kill the video store. Revenge was going to be theirs.

The studios, however, hadn't learned from the unintended consequences they had previously caused. Their next pricing decision led to their own demise from a competitor that they could not see, feel or touch. Like the wheel of a video cassette, what goes around in the movie business seems to eventually come around. This time it would be the online video store and the consumers who would be the big winners, and the studios who would be the big losers.

The biggest own goal since Lee de Forest sold the rights to the triode

DVD collections, the studios thought, would sit neatly up on the shelf alongside people's CD collections. They would therefore not repeat their previous pricing error. They would sell them at a price that competed not with renting a seat at the cinema but with the cost of a rental at the store. After all, why would anybody bother renting a new release for $4.99 when they could buy it for $9.99? The end of the video store was nigh but so too was the birth of a new type of video store.

The arrival of a new technology, DVDs, and the studios' pricing opened up a number of exploitable seams in the video rental industry like the ones that Bezos had so cannily mined in the book industry. This online video store started to look like it might work.

Hacking the video store

The first hack and the one that cast the longest shadow was the solution to the warehouse problem. The warehouse problem, or the tyranny of physical space, which they shared with Amazon, was that an online store that served the needs of a whole country would need a massive amount of storage space (Anderson 2009).

The online video store got around this by using its own customers' houses as the warehouse (Payne 2021). Customers each received three DVDs. Their replacements were shipped only when they were returned. In the meantime, they wouldn't be charged for their rental but instead would be charged a monthly subscription fee. There was no need for a large warehouse since all the discs were at one customer's house or another.

The second hack was how DVDs were replaced. They came from a list of preferred titles that the customer built on the web application. The video store tried to get the customer their top preferences but that rarely happened (Shih et al 2009).

The third hack explains why the customers were happy with their replacements in the exact way that Brendan and I were never happy with whatever garbage Ray sent our way. The online video store knew what their users browsed, knew what they rented and knew what other customers *just like them* liked. In the exact same way that David Shaw had pointed his computers and those who programmed them at problems that had previously eluded such attention, so too did the online video store. They worked on a computer program that recommended movies and it didn't take long for it to get very good.

The 'recommendation engine' that the video store built changed the game in a way that was impossible to predict. First, its recommendations were brilliant. It knew its users' tastes better than they knew themselves. Its recommendations were specific and that created a highly personalised experience. That in turn led to extremely loyal customers. Who wouldn't come back for recommendations that spoke exactly to their tastes? Second, the recommendation engine allowed the online video store to make money – lots of money – on their back catalogue.

The back catalogue in a normal video store serves a specific purpose. Customers are attracted to a wide selection, which they like to browse. But they never rent movies from the back catalogue, preferring almost universally to ignore old movies in favour of the new releases. This meant that in normal video stores, a large selection was required to attract customers but those customers almost always left with the exact same new releases that everybody else wanted to leave with. This exacerbated the new-release problem. High street video stores had to have a wide selection that customers would browse and ignore, as well as all the new releases. The recommendation engine solved the new release problem by giving customers not what they asked for but what they, unbeknown to them, really wanted.

A sliver of personal information, an algorithm and a creative solution to the warehouse problem combined to make this online video store very, very clever.

We'll buy you

Despite how smart these hacks were, however, and despite how much loyalty they engendered, DVDs were still a physical product. The online video store was, in other words, bound by real-world chains that our entrepreneurs couldn't break. That's why they decided to get on a plane and take a meeting in Seattle.

The Columbia Building is in downtown Seattle. It's in

a neighbourhood that's full of strip joints. In 1998, the two entrepreneurs headed to it, stepping over litter and past stores whose windows were smashed to pieces. They made it past the pawn shop, the wig shop that drew in customers from Seattle's transvestite community, the adult entertainment centre and the needle exchange before finally crossing the street and entering into a cluttered, dusty and badly lit lobby (Stone 2014; Randolph 2021).

They were taken through the building where they passed roaming dogs, stained carpets and stairwells that had people working under them. Every surface was covered in paperwork, books, coffee cups and pizza boxes. They were in the offices of a company that made no profit, wasn't going to make any profit anytime soon and had somehow raised $54 million through an initial public offering of its stock.

They heard Jeff Bezos's machine-gun laugh before they made it into his office, where he worked on a desk that, like every other desk in that place, was made from an old door stuck on top of four makeshift legs. Bezos had invited them to Seattle and the headquarters of Amazon.com because it had started becoming the everything store. At that time, books were the only products that Amazon.com sold but that would soon change and DVDs would be a big part of that (Randolph 2021). The quickest way to get into the DVD business was to buy the nascent online video store.

Bezos made a lowball offer that was nevertheless tempting. The video store had a business model that wouldn't scale. DVDs didn't cost $65 but they were expensive to buy and ship and therefore to make any money from, despite the recommendation engine and loyal customers, which in 1998 they'd only just started building. Their destiny, like so many other companies trying to make money on the newly minted World Wide Web, seemed to be the scrap heap. If they knew what was coming in 2002, when the dot-com bubble burst, I suspect they would have taken the offer. It was tempting for another reason. Amazon.com was, one way or another, about to start selling DVDs. They would squish the retail part of the online video store's business like a gnat.

On the flight home, the two entrepreneurs realised they could not go head to head with Amazon in retail. So, they made a brave decision. They would give up selling DVDs directly to their customers. Bezos could have the retail market. They would focus all their energies on DVD rentals. It was this focus that got them through the next few years, through the dot-com crash of 2000, and brought them into my student house in Edinburgh in 2002, where I found three little red envelopes on the coffee table one evening. They were hanging on. Luckily for them, the video cassette wheel of fortune was about to turn again.

Moore's law marches on and information does what information does (again)

Moore's law guaranteed that the chips powering network routers and the computers that sent and received digital signals became increasingly powerful even as their costs went down. This meant that what was happening to computers, the thing that let them go from powering desktop calculators in 1971 to powering desktop computers in 2002, was happening to networks, too – their performance was improving as their costs fell.

We now know the online video store succeeded not by overcoming the insurmountable deficiencies of video stores caused by Universal vs Sony. They succeeded not because of DVDs. They instead succeeded because information and its transformability, along with Gordon Moore's law, was at it again.

If network speeds kept improving, it was just a matter of time before the high bandwidth of the postal system and their trucks could be replaced by even higher bandwidth digital network connections. Once that happened, there would be no need to burn a movie's bitstream onto a DVD and ship it in the post, which was like carrying or shipping punch cards to the mainframe in the computer room. The bits could instead be streamed directly to the customer's home.

This realisation was the spark of insight that changed the

business model: a streaming service would be scalable, profitable and could be consumed by absolutely anybody with a computer and an internet connection. The bottleneck for the video store would no longer be warehouse space or the number of DVDs they owned. The bottleneck would be IT infrastructure.

Movies are re-dematerialised

The arrival of video cassettes transformed the way people interacted with movies. What was once an intangible experience was changed into an object that could be picked up and looked at. The ephemeral became permanent. The immaterial, material. The secret to success on the web, however, was to reverse this trend.

It was the physical elements of web businesses that made them impossible to scale. Pets.com's wonderful website didn't change the fact that dog food needed to be transported from one place to another. What worked for news and music wouldn't work for dog food because Pedigree Chum can't be encoded as bits in a digital stream.

Amazon digitised books with Kindle. The airlines did it with e-tickets. Once the network got fast enough, the video store did it with movies, too. This meant they no longer had to stock as many new releases as possible. They needed only one that could be copied and streamed on demand by one, ten or even a million customers at the same time.

The video store didn't join the pantheon of web businesses on the scrap heap of stupidity. They instead joined a small number, like Flickr, which traded in digital products. The only physical constraint such companies really needed to worry about was IT infrastructure. They needed lots and lots of computers to encode and decode their digital signals, they needed file servers to store the digital copies of the products they sold and they needed cables and routers so that all their machines and databases could be networked together.

In the Noughties, where on earth did our video store entrepreneurs find all this IT infrastructure? They found it at the everything store, of course.

The cloud materialises

In 2007, not long after Amazon had opened up its magical computer store to the public, our video shop entrepreneurs started streaming movies to a small number of users. Not long after, in 2008, they met with disaster. Their database failed. This broke their online video store, which meant that for three long days they could not ship DVDs. They needed an alternative to managing their own IT infrastructure. This brought them back to Amazon, only this time as a potential customer (Izrailevsky et al 2016).

AWS allowed the video store to offload all of its undifferentiated heavy lifting. From that moment on, they could onboard as many new users as they could get because they knew that Amazon and its infinitely scalable cloud could handle them. Almost immediately, streaming movies took over DVD shipments.

As predicted, the magical computer store allowed the video store to use the computer time they needed, thus right-sizing their costs. It allowed them to innovate at pace since that was now dictated by their developers (and marketeers) and not by their access to IT infrastructure. This translated to millions in cost savings, hundreds of millions of users, billions of dollars of revenue and a valuation, in 2023, of almost $200 billion.

Marc Randolph and Reed Hastings had done it. Their online video store, Netflix, had done away with late fees and access to new releases. They caught up with Disney, doing what took them a century to do in only three decades. In 2024, as this book was being finalised, Disney's CEO said that when it comes to streaming, Netflix is the 'gold standard' (Canal 2024). The way movies are consumed really was changed forever.

Amazon and Netflix became essential to each other's success. The magical computer store was the final ingredient, the missing elixir, which would allow the full dematerialisation of the video store. Amazon, on the other hand, had found a famous early customer that proved their computer 'utility' worked. The writing of computing's next chapter had begun.

Where to now?

Following Netflix's lead, companies all around the world tried to use AWS. But they often did something that Netflix never would. They wrapped up access to the cloud in bureaucratic controls, the exact thing it was invented to get around. By doing this, they made sure that their computers in the cloud and those who were supposed to program them sat around idle. This reverse alchemy was how businesses all around the world changed Amazon's multi-billion-dollar computer system into the equivalent of a million-dollar mainframe.

The rare few who did succeed with the cloud were the ones who used it to experimentally develop new features for their web applications and scale their systems, and didn't block access to it like the high priests of mainframes once did.

Like the pioneers of time-sharing who came before them, they saw the cloud not as a utility but as a focal point around which a community of real users gathered. Slowly but surely it dawned on the world of computing that the cloud wasn't a computer utility at all. It was something else entirely.

Chapter 12

A womb with a view

In the 1960s, there was definitely something in the air. Licklider thought the Intergalactic Computer Network would have only a few computers on it and they could form the basis of a utility. At about the same time, the computer scientist John McCarthy, known to his students at MIT as 'Uncle John', envisioned a computer utility forming a new and important industry (Levy 2010). He thought business subscribers to this computer utility would have their own subscribers in the way that Netflix consumes IT infrastructure through its subscription to AWS, while its own customers consume movies through a subscription (Garfinkel 2011). Not long after that, Paul Baran concluded his 1962 paper by asking, 'Is it now time to think about a new public utility for the transmission of data across a large set of subscribers?' (Baran 1962). They were all wrong.

The rise of the grid analogy

The original concept of a computer utility made sense because computers were expensive and had clear societal value. Their development and maintenance costs could be spread across thousands or millions of users and, like the phone and electricity utilities, a computer utility would be a natural monopoly. The concept of a computer as a public utility was a logical conclusion to arrive at, a vision worth pursuing and a seductive argument to add to research proposals.

Capital expenditure

The first argument for a computer utility was all about capital expenditure. Edison's hydroelectric stations were placed next to rivers so that future factories wouldn't have to be. Edison's transformers made sure that electricity was cheap to transmit and safe at the point of consumption. None of this was cheap. Edison and his backers footed the enormous bill for installing the grid. Future entrepreneurs needed less start-up capital because the burden of raising it had fallen onto Edison's shoulders.

What used to be a capital expenditure for businesses was therefore transformed into a monthly operational expenditure based on consumption. In this pay-as-you-go model, a factory that closed for the summer and consumed no electricity would have no bill to pay. Their costs were right-sized to their output and their supply of electricity exactly matched their demand.

Once the cloud arrived, the same was true for companies that built web-scale applications. In the evenings, for example, airlines, with their web-scale booking systems, would need fewer computers and so could safely switch them off. They wouldn't pay the cloud provider in the same way that factories of yesteryear did not pay Edison for electricity they hadn't used.

Democratisation and innovation

The second argument was innovation. In that regard, the cloud does have a lot in common with the grid, both of which have a lot in common with a transformative technology that was invented just a few miles from where I was brought up.

Around 1185, the world's first documented windmill appeared on the River Humber in East Yorkshire. Before the windmill, power came from watermills that were installed on rivers that ran through the lands of manorial lords (Rifkin 2014). This, of course, meant that the lord controlled the power. The windmill, however, could be erected anywhere there was wind, which meant they could be erected everywhere. Just like the grid, not long after the first windmill was constructed, it went viral and was found all over Europe within just a couple of hundred years. That was fast for medieval times.

Because the windmill helped to democratise power, it became known as the 'commoners' mill'. The windmill allowed everyday citizens, who would have otherwise been stuck eking out a living with a plough in a field, to bellow furnaces, mechanically trample wool, mill flour and crush olives. Due to the productivity gains from the conversion of wind into a practical source of energy, these small businesses created products for exchange instead of use. The windmill therefore helped to transform the subsistence economy into a market economy (Rifkin 2014).

In the exact way that the windmill brought forth new small businesses, so too did the grid and later the cloud. In both cases, this led to more inventions and more exchange. An entrepreneur with access, almost as if by magic, to industrial power plants could start up much more easily in the same way that entrepreneurs these days, with similarly magical access to the world's best computer infrastructure, can start up no matter how vast their need for computing 'power' is. John McCarthy predicted this when he said that a computer utility could form the basis of new industries. He was right about that but almost half a century too early. The windmill and grid democratised power. The cloud democratised web-scale computing.

Outsourcing

Finally, as boring as it sounds, the third argument is outsourcing. When Charlie Rose interviewed Jeff Bezos, he gave the example of breweries. Once the grid arrived, the monks in monastic breweries abandoned their old power generator, a furnace. The generator was dirty, dangerous, left furnishings scorched, stank out the brewery and, like a deranged arsonist, threatened to burn the place down at any moment. On top of all that, the generator did absolutely nothing to improve their core product, which was beer (Rose 2009).

Why would any right-minded, beer-brewing Belgian monk waste their time on the infernal generator and its limb-mangling pulleys when they could be working on brewing beer? Edison's engineers were the best in the world. Their craftsmanship was legendary. Even if they wanted to, the monks couldn't have done a better job of producing electricity than the engineers at their local electricity company. It made perfect sense to plug into the grid and let Edison's team do their undifferentiated heavy lifting for them.

The same was true of Amazon's cloud. Their computer infrastructure team was the best in the world. How on earth could any infrastructure team do a better job than the team at the world's largest ecommerce company? They couldn't in the Noughties and they can't now. This is why, in the same way that the grid let the monks focus on their beer, the cloud let Starling Bank focus on their digital bank and Netflix on their recommendation engine and, not long after that, on creating original content.

The fall of the grid analogy

As the years rolled on and time-sharing vanished, the world of computing took a long hiatus from thinking about a computer utility. It started again when David Shaw's protege, permeated as he was with the spirit of ARPA, launched AWS. Not long after that, such is the computer's shapeshifting ability, the usefulness of the grid analogy started to dissolve. Not much more time passed before it stopped making sense altogether. There are three reasons for this.

1. Electricity is not a product

The first is that electricity isn't like any other product or service. It's a dynamic process that's constrained by its machinery so that, at the point of consumption, it's safe. Its users are divorced from its creation, which is usually miles away in a nuclear or hydroelectric power station. The boundary between the vast machinery of the grid and its use within a house or factory is clearly demarcated by the power meter. The user is responsible for the appliances they plug in. Everything on the other side of the meter, such as the dynamos at the power plant, the transformers at the end of the streets, the wires and pylons, is the responsibility of the electricity company.

This is not how the cloud works. When a company uses the cloud, it isn't consuming computer cycles in the same dynamic way that it consumes electricity. Instead, it's renting a computer, which is analogous to renting a generator from the electricity company. It's the customer's job to make sure that they get the most out of that computer. If they can't, that's their problem. No matter how effectively (or ineffectively) the customer uses that computer, the cloud provider will in any case bill them the same amount.

Using the cloud is therefore much more like renting a car. The rental company is responsible for providing a safe car. The customer in turn is responsible for how it's used. If they are to get value for money from it, they will have to drive it. The rental company doesn't care if the customer drives the car or parks it on the street. This makes the cloud vastly different to the grid, the 'product' of which is a flow of electricity that's paid for by the unit and measured by a meter.

2. Underperforming capital assets

The second reason is to do with capital assets and how they should be used. Not long after the first electricity grids came online, popping up in cities where the companies were called Edison Boston, Edison Chicago and so on, metering became more accurate and thus gave the power companies more insights into how their customers consumed electricity. Not surprisingly, department stores didn't use much in the evenings, factories didn't use much during lunchtimes and families didn't use much during the day.

This gave the power companies a headache. Extremely expensive generators, depending on the time of day, were an underperforming capital investment (Schewe 2007). This made it hard for electricity companies to 'sweat the asset'. Nobody back then thought that was the customer's problem. It was instead patently obvious that it was the power company's job to make sure all the machinery was highly utilised, hence the headache.

A well-sweated asset drives revenues up and costs down. This is why, in one particularly creative campaign, free electric irons were given away to households that signed up for electricity; the extra consumption would lead to increased usage and give the power companies a chance to sweat their assets more effectively. For the same reasons, when it still shipped DVDs, Netflix gave its customers a month's free subscription knowing that, once they tried its video store with its recommendation engine, they would fall in love with it. Netflix's DVDs were the assets that needed sweating.

Cloud providers also want their assets to be highly utilised. They don't want to add more computers to their data centres when they can be much more creative, for example by making sure that physical machines are used by multiple customers at the same time. However, once a customer, whether they're a business or a student in their dorm room, has rented a computer from the cloud provider, the high utilisation of that asset, the responsibility for sweating it, is theirs. This is the exact opposite of how the grid works.

3. Natural monopolies and their regulation

The third reason, and maybe the most important one of all, is that the cloud is unregulated. At the turn of the last century, it wasn't obvious to everybody but it soon became clear that the supply of electricity was going to be more effective if it was organised as a natural monopoly. Why duplicate wires, generators, dynamos, sockets and feeder cables? If that happened, then lots of assets would have sat around doing nothing and none of them would ever be sweated. The waste would have to be passed on to customers as higher costs. In that case, electricity would remain a luxury and, without economies of scale, those power companies would in turn fail.

The problem was that the monopoly-busting zeal of the US government, led at that time by Theodore Roosevelt, wouldn't stand for monopolistic practices. With this in mind, and foreshadowing Ted Vail by years, Edison's protege Samuel Insull argued that the only way to run a private monopoly was for it to be publicly regulated (Schewe 2007).

This is why. Natural monopolies emerge if they're essential to society's functioning. They often have high start-up costs, which prevent competitors from entering the market. This in turn makes it more efficient for a very small number of organisations to meet society's needs. It simply doesn't make sense for lots of small power companies to independently provide electricity, not least because there isn't enough sidewalk to bury more than a couple of sets of cables.

First movers therefore have huge advantages. This was true for Edison and the Bell Telephone Company in the same way that it was true for Amazon, Google and Microsoft, all of whom moved quickly to provide cloud computing. Their dominance, like Bell's former dominance, means it no longer makes sense for any other company to try to compete, because they can never catch up and, besides, competition would lead to worse service for users.

Because they're good for society and have an unfair advantage, natural monopolies have to be either owned by the state or heavily regulated. That's true for the electricity grid and it was true for Bell, which licensed its transistors cheaply because Uncle Sam was ever ready to slap AT&T with an antitrust suit. That's not true for the cloud, despite the fact that cloud computing, which has just a handful of players who moved first and took advantage of their scale, is without a doubt a natural monopoly.

Shared responsibility

The main problem behind the grid analogy is that it hides or plays down a key feature of the cloud. Cloud computing is based on a shared responsibility model. The cloud provider is responsible for computer infrastructure and the buildings it's housed in. The customer is responsible for the applications that are running in it. Managing these applications is much more complex than managing the applications in a house. The 'grid' team in my house is made up of one person (me) and my process involves walking around after my kids, switching off the lights and, while I'm at it, my wife's hair straighteners.

A company with just 20 web-scale applications running in the cloud would need a whole team to manage them. That cloud team would be responsible for data integrity, local regulations, compliance and the utilisation of the computers they had rented, to name just a few things.

There's no shared responsibility in the grid. The responsibility for applications running in the cloud falls on the customer, whereas the responsibility for the computers and their housing falls to the cloud provider. Any manager who convinces themselves that the cloud is comparable to the grid is therefore in for a nasty shock, pun fully intended.

Remarkably, businesses either aren't aware that cloud computing is based on a shared responsibility model or they discount it, convincing themselves that managing applications in the cloud is a bit like switching off hair straighteners.

A poor analogy

Was the grid analogy a cynical marketing ploy? Was it used to disarm unsuspecting businesses by making them believe it was safe, easy to use and regulated? It would make for a great conspiracy theory, and it's not totally false. But it's also not totally true.

The grid analogy was seductive in the 2010s for the same reason that it was seductive in the 1960s. It made sense. The cloud was a commoner's computer, Amazon did take the hit on billions of dollars' worth of capital so other businesses didn't have to, and that did mean they could focus on their product while Amazon did their undifferentiated heavy lifting for them.

However, once the monthly cloud bill arrived, like it once did for Bill Gates and the Lakeside kids, people started to realise that the cloud is built around a shared responsibility model and therefore, on second thoughts, that it isn't much like the grid at all.

Where to now?

American author and teacher Joseph Campbell once said that marsupials, such as kangaroos, spend the first part of their life in a 'womb with a view'. It's from the safety of this second womb, the pouch, that joeys begin to understand the world around them (Campbell 1995).

Myths fulfil a similar purpose for humans. We learn about life, death, love, loss and what happens when we disobey our parents from stories about little girls in the woods and the big bad wolves that eat them. At some point, like the kangaroo, we climb out of the safety of the womb ready to face the world as it really is.

It's a myth that the cloud is like the grid. Some would say it was a benign myth that gave companies the courage to start on their journey like Perseus's winged sandals and shield gave him the courage to start his.

Others would say it was a self-serving myth propagated by cynical salespeople that convoluted the highly reliable and regulated electricity grid with warehouses full of computers that have nothing to do with carefree and fluffy clouds and instead have everything to do with collecting personal data, building artificial intelligence models and doing so while driving profitability through the roof and, where possible, avoiding paying any tax whatsoever. Either way, it's a myth. But it is a myth that we have now fully dismantled.

We're climbing out from our womb with a view but that, of course, leaves us with a nagging, unanswered question. If the cloud isn't a computer utility, then what is it?

Chapter 13

The computer as a community

Time-sharing computers taught us two lessons. The first was that, once in users' hands, computers are used for purposes their creators almost always failed to foresee. The second was that in a computer 'grid', 'power' flows both ways (Waldrop 2018).

Get it in a user's hands

The creators of the electricity grid had no idea which applications humanity would go on to use it for. They didn't predict that factories would soon be illuminated for 24 hours a day and from their incessant work, a globalised textile industry would arise. Edison had a hunch, hence the specific wording in his patent application, but even he didn't envisage that a huge array of converters, from heaters to radios, televisions and later electronic computers, would be added to the light bulb to create an array of applications for electricity and the grid that supplied it.

The same happened with the invention of the transistor. Beyond using it in a hearing aid, the scientists at Bell Labs had no idea of any potential commercial applications for their fabulous midget. However, they did know that in the hands of real users, the transistor would be woven into applications that were at that time unimaginable, even to them. This is why, encouraged by Uncle Sam, they arranged lectures, wrote up educational materials, provided free samples to companies such as Motorola and RCA and later licensed the transistor for next to nothing to companies like Sony.

One such unimaginable application was the microchip. It was soon found careering around in the nose cones of ballistic missiles and spaceships less than 50 years after the Wright brothers first flew in an aeroplane.

This process repeated itself again when the microprocessor was invented in 1971. Intel could not see beyond Busicom's calculators and specific applications that would need control logic, such as Bill Gates's Traf-O-Data system or the new-fangled electronic pinball machines that started appearing in honky-tonks and seaside arcades.

That changed when Intel provided samples of microprocessors to potential users. All kinds of applications, including the semiconductor laser, were subsequently created and their destiny was to power the whole internet. The microprocessor was also built into Ed Roberts's personal computer, which itself went on to birth a mind-boggling array of software applications and industries.

The same happened to time-sharing computers. Once in the hands of real users, time-sharing evolved in a way that absolutely nobody saw coming.

Time-sharing and the first hackers

MIT was part of Licklider's Intergalactic Network and therefore received funding from ARPA. Some of that funding was spent in 1959 by John McCarthy and Nat Rochester to develop and launch MIT's first programming course (Waldrop 2018). Their class contained several members of the Tech Model Railway Club (TMRC), whose hobby was to build train tracks with electrical and telephone switches that they scavenged from the local junkyard (Levy 2010). At first glance, it may seem strange that model railway enthusiasts would be into computer programming, but these two things are more related than you may dare to think.

Railways had, for years, used machinery to make sure that no trains would ever be sent towards each other along the same single piece of track. If a signal was green for one train then it would be red for the other. The mechanical construction of model railways, with their if–else conditions and tracks that went around in loops, was not just similar to the Boolean algebra that underpins all computer programs. It was almost identical to it. The logic 'gate' that makes sure trains don't crash on the same piece of track is similar to what computer scientists call an 'exclusive or' (XOR). If you've ever walked into a room and turned a light on only to leave through another door where you turned the light off, then you know what an XOR is. If both switches are off, the light will be off. If both switches are on, the light will be off. Only if one switch is on and other is off will the light be on. That's XOR. It stopped trains crashing long before us computer wonks adopted it.

The model railway club had its own rules and culture. It valued ingenuity and mischief. A piece of kit that wasn't working was 'losing'. If it couldn't be saved, it was 'mashed up until no good' – *munged*. Any student who wasted time going to class or doing the actual work their professors set for them, instead of messing around with trains, was called a 'tool'. Finally, they called ingenious track designs 'hacks', a word already in use at MIT for ingenious pranks. It didn't take them long to use their new programming skills to pull off the granddaddy of all computing hacks.

The model railway hackers found a transistorised computer, the TX-0, on the first floor of the computer centre at MIT. The TX-0 was a multi-million-dollar machine that used transistors instead of vacuum tubes. Transistorised computers were like horseless carriages or the wireless telegraph, the original name for the radio. The replacement of a single component didn't change their overall design but radically altered their performance (Ceruzzi 2012).

Alone with the TX-0, the hackers got busy. When they returned to the computer room, members of the faculty were horrified to discover that some idiot had used the $3 million TX-0 to re-create the functionality of a $300 typewriter. They'd saved their program in a file called Expensive Typewriter. Salt was rubbed further into the wounds of the old timers' indignity when, on another trip to the computer room, they found a program that did basic arithmetic. This one was called Expensive Calculator (Levy 2010; Waldrop 2018).

The conflict

Ever since it was invented, there have been two viewpoints as to what the mechanical computer is and how it should be used. These different viewpoints separate their holders into mutually antagonistic groups.

The first group sees the computer as an expensive aid to mathematics. They value the time of the machine more than they do the person programming it. This group holds the machine in reverence and therefore protects it behind customs and rules. It's in the space between reverence and rules that priesthoods and superstition emerge.

The other group values the programmer's time. They see the computer as an aid to creativity. In their eyes, it's a machine that allows the user to interact with information and knowledge and the process of doing so is exhilarating. The user never wants to stop. This group holds the process of interacting with the computer in reverence.

The old timers at MIT were part of the first group. They thought that programs should be developed carefully and in advance, so as not to waste the machine's time. They viewed the computer as a big and expensive calculating machine that accepted complex sums as input and produced answers as their output.

The Young Turks of the model railway club and their peers, as members of the other group, saw the machine as an aid to creativity. They thought that programs should be developed and refined experimentally through a process of trial and error, exactly how Expensive Typewriter and Expensive Calculator were. Unsurprisingly, the old timers thought that using the machine as an aid to creativity was a ludicrous waste of machine time. The Young Turks, also unsurprisingly, thought that fully working out a program in advance without trying it out on the computer was a ludicrous waste of human time.

The power struggle between the old timers and Young Turks at MIT was not superficial. It was emblematic of a much deeper and often subconscious disagreement. The real question was about how computers should be used and accessed. Was the computer an extension of our human selves, like a chair or an abacus? Was it, in other words, just a tool to assist other fields of study? Or was the computer an interactive device that took instructions and gave feedback in real time, like a car? Was it, in other words, a machine with which humans could form a symbiotic relationship?

Time-sharing communities

What the hackers at MIT showed, and what the creators of time-sharing didn't see coming, was that users' programs became part of the system itself. Once a time-sharing system was online, it would evolve at a pace dictated not by its designers but by its users. This is how, in the 1960s, programs such as MAIL, which allowed users to send each other text messages, and ARCHIVE, which compressed files, were created by some users and made available to all the others.

Unexpectedly to their designers, early time-sharing systems quickly stopped being 'grids' of computing 'power' and almost immediately became the focal point for a community of real users who could message each other, swap files, upload their programs for others to use, argue about their improvements and, of course, gossip. The more users a time-sharing system had, the more valuable it became; new programs would attract new users who in turn would write new programs.

In other words, time-sharing computers and the communities that arose around them looked very much like the sort of communities that would emerge, on a much larger scale, across the web in the 1990s. It was this community of users that shaped Licklider's thinking when he dreamed up the Intergalactic Computer Network. A great lesson had been learned.

The cloud as a computer utility

When Netflix launched, they had to build their own web pages and provide a way for customers to pay, log in and browse films. Marc Randolph said that companies back then had to get investors and early employees to bet on their idea (2021).

Once their investment was raised, a company had to build a team who then had to build their application, the IT infrastructure on which it ran and any auxiliary software that other companies might one day need but that hadn't yet been invented, such as a recommendation engine. This was the undifferentiated heavy lifting that all web-scale companies had to do despite it not helping them with their core product. The cloud fundamentally changed this.

AWS's first services were storage and computation, but over time Amazon released new services. For example, in 2009, auto-scaling functionality was added to AWS. This allowed users to automatically scale their usage up and down, making it easier to right-size just how much computing power they paid for.

Once AWS was available to the public, Amazon had brought its vision to life. Businesses and students in their dorms really could easily build web-scale applications on top of the world's most successful ecommerce company's IT infrastructure.

What its users couldn't do was upload that web-scale application or other programs like the TMRC hackers had uploaded Expensive Calculator to the TX-0. For a brief moment, in its IaaS phase, AWS therefore looked a bit like a computer utility.

The cloud as a community

The overriding design principle behind AWS was that all web-scale applications needed the same handful of building blocks. They needed a way to collect payments, store files and, of course, they needed computation to process all the bits floating around in their digital streams. From these computing 'primitives', complex and unforeseeable applications arose in the same way that simple building blocks give rise to life.

Similarly, the designers of the first time-sharing machines made sure that basic programs, for example to write and read data to their magnetic disks, were available to all users. These fundamental programs allowed new users to write their own 'higher-level' programs without having to read the manual, which is how the model railway hackers had been able to write Expensive Typewriter so quickly. In the same way, AWS's customers could build web-scale applications without worrying too much about the low-level functioning of databases and file servers.

In the 2010s, once Amazon had sorted out the building blocks, they made it easy for others to upload their programs to the cloud in the same way that it was easy to upload programs to time-sharing computers. Amazon called their new system Marketplace. Their 2012 press release said that Marketplace was an online store where its customers could easily find and deploy software that runs on AWS.

This was, once again, John McCarthy's vision made manifest. Users would not just use AWS directly but could use software from Red Hat, IBM or indeed any company that wanted to write an application and share it like Expensive Calculator had once been shared.

Once Marketplace arrived, it wasn't just computer infrastructure that was available as a service (IaaS) but software, too. This was called Software as a Service (SaaS). If Netflix were starting out today, it could use ElectrifAi, which is available on Marketplace, as its recommendation engine. It would cost a few dollars an hour. This would let Netflix focus on what it does best, which is streaming and making movies, while the team at ElectrifAi focus on what they do best, building recommendation engines and therefore putting people like Ray out of a job.

Marketplace allowed companies to focus on the solution they were building rather than the IT infrastructure or the auxiliary software they needed. This in turn removed some of the speculation that investors had to do; they could now make decisions based on preliminary results and the entrepreneurs would get to keep more of the businesses they founded. At the same time, those without capital were no longer excluded from starting a business. SaaS therefore changed the software and the investment games at the same time. The cloud as a community had arrived and the grid analogy no longer made any sense.

No clear boundaries

If the history of computing teaches us anything, it's that there are no clear boundaries. Instead, ideas leak into each other. Time-sharing bled into the work at ARPA, which bled into the work at PARC. This is partly because of the pace of change, which in any case was so quick that the heady early days, for example at Amazon and Menlo Park, quickly atrophied into legend. It's partly because so often there's just something in the air – systems are built from technologies and new ideas that are often already floating about. Partly, it's because so much money is at stake that it serves people to muddle or rewrite history. However, it's also because the computer shifts its shape in ways that can't be predicted while shifting the shape of the things it touches, changing what it means to be a bookshop, a video store and a camera, to name just a few examples. Computers dissolve boundaries, *including their own.*

The same was true for the cloud. It wasn't exactly the case that before Marketplace there was only IaaS and after it there was SaaS. Programmers have been sharing their work with each other forever, either on disks, through open-source projects or even on scraps of paper. Salesforce provided their software as a service as early as 1999. Their application was available over the web using a browser. In the same year, a company called LoudCloud was created and it let users host their web applications.

With that said, however, neither LoudCloud nor Salesforce can claim they invented SaaS. In the 1980s, airline software was provided as a service on mainframes and travel agents dialled in with terminals from high street shops. If you remember your grandma's radio, then you might remember those green screens of holiday magic, too.

Marketplace was, in other words, the continuation of an existing trend. It was a trend first spotted in the communities that sprang up around time-sharing computers and later became entrenched across the web as thousands of programmers joined open-source communities and bulletin boards where they could swap ideas and, importantly, software.

What Marketplace did, though, was make it easier, especially for businesses, whether they were starting up or kicking off new initiatives, to cobble together software and launch new products and gather feedback quicker than ever before. Marc Randolph could only dream of such a system in 1996.

In his 2021 memoir, *That Will Never Work*, Randolph recounts a drive from Scotts Valley to San Francisco. His son was riding shotgun. On the drive, as Randolph did the driving, he managed to build a web app, set up some analytics to monitor its usage, used a payment tool called Stripe and even managed to buy some adverts. Scotts Valley is one hour's drive from San Francisco. Back in 1997, Randolph said this would have taken six months, a whole team of people and therefore a million bucks to do (Randolph 2021). A lot changed in the two-and-a-bit decades since he started Netflix and his son was hacking on his laptop in the car that day.

Where to now?

In the hands of Netflix's engineers, Amazon's cloud was shown to be a devastatingly effective engine of experimentation. In less capable hands, it was just devastating. Not long after the cloud arrived, a pattern emerged. It turned out that those wielding the cloud, and their relationship to this weapon of mass creativity, were as important as the cloud itself.

Chapter 14

Talent density

The cloud was so powerful and, as Amazon and Netflix had shown, so disruptive, that the early 2010s seemed like a perfectly rational moment for big businesses to panic. So, panic they did.

Three-way threat

None were more jittery than the banks. They were threatened in three ways. First of all, so-called 'challenger' banks could get themselves an AWS account and before the old banks knew it, their customers would be closing their accounts and opening up new, digital-first, mobile-only accounts.

This was already a threat during AWS's early years and was therefore an even bigger one once Marketplace had launched and open-source tools such as Kubernetes and Docker matured and brought with them the sort of competitive advantages that were previously only available to companies like Google. Cloud-first challengers, it was feared, would do to the banks what Netflix was in the process of doing to Blockbuster.

The second threat came from the technology companies themselves. Amazon had already annihilated high-street bookstores and specialised shops such as Toys 'R' Us. This led, at the turn of the century, to Amazon joining Xerox, Hoover and Google in becoming a verb; companies lived under the constant threat of being Amazoned.

What if Amazon moved into finance and Amazoned all the banks? If not Amazon, what was stopping Facebook? They had so many users and so much of their personal data that offering banking services seemed like an inevitability.

The final threat came from other banks. They were all scared of each other. The bank that started using the cloud first would outmanoeuvre the others and in doing so win the war for digital users and doom the rest to perpetually play catch-up.

This last threat was not lost on sales teams at the cloud providers. They stoked and then preyed on this fear. This was maybe not surprising, since feeding on the paranoia of managers was, quite literally, the oldest trick in the automation salesperson's book. It had worked for those peddling looms and gig mills in the time of the Luddites, it worked during times of factory automation in the last century and it worked in exactly the same way when the cloud arrived in the late Noughties (Merchant 2023).

The Netflix of the Netherlands

It was at this time and within this context that I found myself catapulted into the peculiar world of banking. In 2015, I was invited to a Dutch bank that was determined to fight off threats from the technology companies and start-ups so they could become the incumbent that would batter (rather than be battered by) the competition. They were panicking in their own way, slowly.

When I arrived, there was definitely something in the air. The whole place was electrified with ideas about modern software and IT infrastructure practices. They had become obsessed with how they would solve the IT infrastructure bottleneck so they could develop web-scale applications as quickly as Amazon and Netflix developed theirs.

The bank's ideas came from the blogs and books their team read and the conferences they had attended. This is how they learned about techniques like chaos engineering from Netflix's engineers. It's also how they discovered how Netflix used AWS to scale their cloud computing usage up and down depending on how many people were watching movies. If Netflix could do all that, why couldn't a bank?

After touring the engineering teams and seeing their monitors and whiteboards, we grabbed a coffee and ducked into an empty office. It was here that Frank (not his real name), who was showing me around, told me that the bank had developed a crystal-clear vision that would not only make them the coolest place in finance to work, but would give them the tools they needed to beat their competitors in the nascent battle for web and mobile users.

Frank leaned forward in his chair, so I did the same. We came face to face. Looking directly at me, he flicked all his fingers off his thumbs to reveal two massive open hands and said, 'We're going to become the Netflix of the Netherlands.' He paused before throwing his gigantic frame back into his chair, where he tapped his feet, wiggled his fingers and giggled like a child.

The complexity question

The problem for Frank, and I really wish that it hadn't fallen to me to tell him, was that he worked for a bank and, you know, Netflix is a video store. It was a cool video store. It had upended the industry. But it was a video store nonetheless.

Frank assured me that what worked for Netflix would work for his bank. I didn't totally disagree. But the systems required to run a bank were vastly more complex than anything Netflix was building. Banks are also highly regulated. You could not, especially back then, host customer data in a foreign country. What was possible for an online video store, unencumbered by decades of legacy systems, policies and thinking, was impossible for a bank. Frank was either playing down the challenge ahead or was deluded.

The workforce question

I asked Frank how he was going to get his teams experimenting. This drew a puzzled look that, as we continued, developed into a scowl. Frank was royally pissed off with me. I changed tack. I asked about roles. Adopting cloud computing means that some roles will no longer be relevant. The people who look after physical computers, for example, will no longer have a job because that work will be outsourced to the cloud provider, in the same way that the old job of looking after a power generator was outsourced to Edison Electric. Some roles (and those who will fill them) will be brand new. These roles will be about monitoring cloud costs, managing compliance rules and making sure that data is secure.

This line of questioning matters, because cloud computing simultaneously creates new roles within a company while making many more redundant. This is why adopting cloud computing and workforce management go hand in hand. It is, unfortunately, something that most managers and the businesses they work for cannot stomach. More frequently, they haven't thought about it at all, or they've discounted the impact of cloud computing on the workforce, which is what Frank did.

Once I shifted to asking about roles and responsibilities, the meeting with Frank ended. The bank he worked for then spent hundreds of millions of euros trying to become the Netflix of the Netherlands. They failed.

In places like Amazon and Netflix, the cloud is put to work in the service of experimentation. This requires engineers and managers who are curious, risk seeking and serious about learning from failure. Those drawn to work at large companies such as banks or pension funds rarely have these characteristics. I would have told Frank this but he had already thrown me out.

Cloud computing places companies in a predicament. On the one hand, they desperately need the cloud to compete. On the other hand, they don't have the capabilities or the mindset to succeed with it. The only way to develop the capability and the mindset is to refresh their workforce. This was the real lesson that Frank and his colleagues failed to learn from Netflix.

The road to Damascus

In his book about Netflix, *No Rules Rules* (2020), which he wrote with Erin Myer, Reed Hastings, the co-founder and second CEO after Randolph, recounts a story from the spring of 2001, when the internet bubble burst and all of Netflix's funding dried up.

Hastings had the unenviable task of letting go of about a third of Netflix's 120-strong workforce. He couldn't see who was failing in their job and, in fact, many were not. Together with Randolph and Patty McCord, Netflix's chief talent officer, Hastings started to divide people into two groups. Anyone who could collaborate, did great work and was creative went into their keepers' pile. Others, who worked hard but moaned or were high performers but didn't collaborate well, would be made redundant.

After the redundancies, Hastings, Randolph and McCord braced themselves for a backlash that never came. In the following days, the much smaller team was both happier and more productive. Of this experience, Hastings said, 'I discovered something that completely changed the way I understand both employee motivation and leadership responsibility. This was my road to Damascus experience, a turning point in my understanding of the role of talent density in organisations. The lessons we learned became the foundation that led to Netflix's success' (Hastings & Myer 2022).

Talent density

What did Hastings mean by talent 'density'? Talent to Netflix meant creativity, collaboration, competence, good judgement and optimism. Before the layoffs, the talent of the 80 was 'diluted' by the 40 who didn't share these values (Hastings & Myer 2022). This is why, after the layoffs, Hastings said the talent of the organisation was the same but was not watered down by those who didn't share these values. This is what he meant by talent 'density'. This idea formed the basis of management at Netflix (as it once formed the basis of management at Menlo Park, in the deserts of New Mexico and then later at Xerox PARC). When Netflix's attitude towards talent and their system of management collided with an infinitely scalable computer utility, the results were startling.

Predicting success

Now that we have a better idea of what was happening at Netflix in the 1990s and Noughties, we can be more confident when it comes to predicting success or failure with the cloud. There are four questions that all companies moving to the cloud need to ask and answer. If the answer to them all is yes, then success is not exactly guaranteed but at least the probability will be skewed in their favour.

1. Does the company see the cloud as a system of innovation?

If a company or its managers think that the cloud is like an energy utility from which they can draw computer 'power', then there's a good chance that they're desperately looking for a magical solution to their problems. They're likely to wrap this magical solution up in rules and regulations and thus re-create the sort of priesthood that the cloud was invented to do away with in the first place. If, on the other hand, they see the cloud as a systematic aid to creativity, then there's a good chance that they understand how companies such as Netflix use it.

2. Is talent valued?

This is just another way of asking, 'Does the company think success in business is driven by people?' As bizarre as this question sounds, many companies and the managers that run them don't understand the importance of a talented and motivated workforce. To them, people are interchangeable parts that can be swapped into and out of the 'machine' of work. Nobody in the history of computing made this mistake. Edison didn't make it. Oppenheimer didn't make it. Licklider definitely didn't make it and neither did Taylor, Bezos or Hastings.

3. Is talent defined?

If a company believes that success with the cloud comes not from technology but from the talent of those wielding it, then they're heading in the right direction. However, they have to not only value talent but also have to sit down and define what they value and therefore what talent means to them. If values and talent aren't defined, then they can't be embedded into HR processes, job specifications and competency models, which is where the hard work begins.

4. Has the definition of talent been embedded into the company's processes?

Once a company has defined talent, they have to embed it into their hiring, promotion and dismissal practices and processes. The core values of my company, Container Solutions, are growth, trust, purpose, collaboration and creativity. Each part of our hiring process will test for one or more of these values. When a candidate makes it through to the end of the process, we can say with some confidence whether or not they share our values, which we believe are essential to success with cloud computing and systems development.

Similarly, performance review processes also have to be based on values. In good teams, a positive review, and therefore a promotion, will only happen when people are doing their jobs well and living the company's values.

Finally, management must have the skills, courage and supporting culture, processes and practices to remove employees if it's not working out. It may not be working out because of a mistake in hiring or it may not be working out because an employee and the company have grown apart. This happens when companies adopt the cloud because certain roles are no longer required.

One way that Netflix deals with the problem of removing employees is with the 'keeper test'. The keeper test asks managers to ask themselves, if a person were to quit tomorrow, would you fight to keep them? Or would you feel relieved? If it's the latter, offer them a severance package and start looking for somebody else to fill the role (Hastings & Myer 2022). The keeper test shows just how seriously Netflix takes talent management.

Systematic leadership failure

Netflix was always clear that talent density and a culture that valued freedom and responsibility formed the foundation of their success. As early as 2009, Hastings shared a PowerPoint presentation called the Netflix Culture Deck. It has been downloaded tens of millions of times and was described as one of the 'most important documents ever to come out of Silicon Valley' (McCord 2014). It was (and still is) an outstanding management guide for companies trying to succeed with web-scale software and cloud infrastructure development. Many of the same management tips reappear in *No Rules Rules*.

Why do so many managers read and then discount Hastings's deck and book while at the same time remaining steadfast about the technologies they want to use? In his fantastic book, *Why Do So Many Incompetent Men Become Leaders?* (2019), Tomas Chamorro-Premuzic goes a long way to answering this question. Human beings find it hard to tell the difference between confidence and competence. Often masked by charisma, arrogance is almost universally mistaken for leadership potential. This only happens to men because when women behave the same way, they're judged differently. Women and humble leaders of any gender are therefore not considered for leadership positions and organisations are thus left in the care of incompetent men (Chamorro-Premuzic 2019).

This scuppers success with the cloud. The ability to build and maintain high-performing teams, like McCord and Hastings did, and to get people to set aside their own egos and ambitions for the group, as Bob Taylor did at PARC, are both vital to success, because web-scale and cloud infrastructure development both require multidisciplinary teams. Arrogant men can't help others to set aside their egos because they can't set aside their own. An arrogant leader will read the Netflix Culture Deck and immediately discount or ignore it because they think it doesn't apply to them.

Such 'leaders' tend to avoid risk until their reputation is on the line. It's at this point that they become not risk-seeking but reckless (Dixon 1994). In other words, they resist the inevitable course of action until the moment they're forced to act. Then, like a character in a William Faulkner novel, that one with Dewey Dell, they make out that it was always their intention to get moving in the direction that circumstances are now dictating (Faulkner 1930). Allowing events to dictate a course of action is the opposite of leadership.

So, what gets a person promoted into a position where they have to lead their organisations into the brave new world of cloud computing is the exact same set of characteristics that disqualify them from the job. Because leaders like this are so full of fear, and because events have caught up with them, wishful thinking, that great protector of the ego and shield to anxiety, allows them to accept that the cloud is just like using the electricity grid and, of course, just as easy.

Does this also explain why the same leaders discount the management lessons of Netflix but cling on to the technical lessons? Sort of. Management is really hard. Technology is really cool. Individuals aren't drawn to technology because they want the responsibility of managing other people. The opposite is often true: they're drawn to technology because it lets them avoid other people. That was certainly one of the benefits of the time I spent with the Commodore 64. Incompetent managers want money, status, power and fun – and technology is fun. They don't want any real responsibility and certainly don't want to be held to account. This is how they commit their organisations to years of misery, burning out their people and blowing tens of millions of dollars in the process.

The IC problem

In the world of computers and programming, those who do work, rather than those who manage others, are called individual contributors (ICs). ICs are the other reason why companies are full of incompetent leaders. ICs are judged through their results and their social standing within a group. Leadership, on the other hand, requires different skills as well as emotional maturity and especially a lack of narcissism. Does anybody really think that any of the individual nutcases that Bob Taylor managed at PARC could, at that point in their lives, manage other people?

The problem is that ICs get promoted into management positions where they're doomed to fail. Bersin and Chamorro-Premuzic say that the IC problem explains why the world's best athletes can become mediocre coaches and why mediocre athletes can go on to be great coaches (Bersin & Chamorro-Premuzic 2019).

Where to now?

In the early days of Netflix, Patty McCord and Reed Hastings used to carpool to work. One day, on the drive to work, after those layoffs, when Netflix's new talent density had boosted their energy and creativity, McCord asked Hastings what was going on. She wanted to know if she was in love or if she was under the effect of 'wacky chemicals' that would soon wear off (Hastings & Myer 2022). McCord was describing what psychologists call a 'peak experience'. Without knowing it, the team at Netflix had stumbled into the world of Abraham Maslow.

Chapter 15

The computer and the company

In 1945, smack bang in the middle of J C R Licklider's time at Harvard, which ended in 1950 when he left for MIT, one of the most famous and controversial figures of 20th-century psychology took up space on the other side of the corridor to Licklider's psychoacoustic lab.

Skinner

B F Skinner was born and raised in the railway town of Susquehanna. Its rhythms were set by the train timetable and blasts of the repair shops' whistle, which indicated when shifts at work started and ended, when shops opened and closed and when meals were served (Buxton-Cope 2020).

When he was a boy, with the repair shop in view, Skinner taught himself to build experimental equipment through a process of trial and error. Like Edison, his ambition was matched by his work ethic. The results back then were see-saws, guns and a glider that refused to fly.

Skinner's talent for tinkering, his observations of the citizens of Susquehanna responding to the whistles of the train yard and his work ethic reappeared in his work when he first landed at Harvard as a postgraduate student under William Crozier.

Crozier's approach to psychology focused on understanding behaviour without worrying about the workings of the mind. It was during this time with Crozier that Skinner invented his famous 'box' that rewarded rats if they managed to push a lever. This work led to Skinner's first book, *The Behaviour of Organisms*, which he published in 1938, after graduating and moving on from Harvard.

When he returned to Harvard and Licklider's corridor, Skinner was the leader of a movement that had hardened at first into dogma but by 1945 had atrophied into something more akin to a religion. In Skinner's hands, behaviourism had become radical behaviourism.

Radical behaviourism

Radical behaviourism forbade its adherents to look at mental states such as happiness and fear. Along with purpose and consciousness, these states were off limits partly because they couldn't be observed but mainly because Skinner thought that they, like the mind itself, did not exist.

To the radical behaviourists, consciousness was an illusion. Human brains, instead, did nothing more than change stimuli into responses. The mind was a black box whose contents were not only unknown but unknowable. Inputs went into the box like punch cards went into old computers and looms, and in response to them, outputs came out.

Mechanical life

At the other end of the corridor to Skinner's, Licklider and his colleagues thought that humans had minds, not least because at that moment they were mulling over a radical idea of their own.

During the war, it had been almost impossible for gunners to shoot down German bombers or fighters. As I discussed earlier, the gunner had to fire the shell into the space in which they thought the German plane would be. This was a computation that a machine might be able to make but a human, even with the most up-to-date firing tables, rarely could. Like computers, the German planes and their zig-zagging were, nevertheless, bound by the laws of physics.

The flight patterns of the planes were therefore predictable.

During the war, Licklider's colleague, Norbert Wiener, devised a mathematical model that predicted where those zig-zagging planes were going to be (Haigh & Ceruzzi 2021). A gun, connected to a computing machine embodying this model, could fire a shell so that it arrived in the sky at the same moment the plane did. The computing machine would calculate in real time. No firing tables were needed, nor was a human. The machine pulled its own trigger.

Wiener's machine was trying to achieve a goal in the future. The implications were profound. Wiener's gun taught Licklider and those around him that, when built with a system of feedback, machines embodied purpose and therefore a rudimentary sense of mind.

In 1945, Licklider and those in his lab mulled over ideas about mechanical life at the same time and on the same corridor that Skinner and those in his lab were convinced that human beings were nothing more than organic machines.

Licklider thought there had to be a third way to think about the human condition, one that was neither mystical on the one hand nor in denial of the mind and consciousness on the other (Waldrop 2018). If a machine had a mind, he thought, then so too did a human being.

Luckily for us, Licklider was not the only one having this thought in 1945.

The third force

Sigmund Freud's psychoanalysis framed human drives and motivation through the lens of inner conflict and instinctual urges. This was the first force in psychology. Skinner's behaviourism was a reaction to this. Its focus, in contrast to psychoanalysis, was on observable human behaviours, which were shaped by consequences and reinforced by repetition. Humans never act; they only react. Behaviourism was the second force in psychology.

Humanistic psychology, which is what Licklider was sensing in the mid-1940s, was a common-sense reaction to radical behaviourism. It wasn't fatalistic – humans weren't bound by their animalistic urges. Nor was it manipulative, in the way that behaviourism and its focus on perfectibility through the control of the environment was. Instead, humanistic psychology had a hopeful, growth-focused perspective on the human condition. Its practitioners considered the person to be whole, to have agency and their own view of the world. It stood in defiance of radical behaviourism. Humanistic psychology became the third force in psychology.

Abraham Maslow

The architect of third-force psychology was Abraham Maslow. Maslow was interested in the relationship between creativity, self-actualisation (a concept he invented), meaning, motivation and religious or transcendent experiences.

In 1961, the *Journal of Humanistic Psychology* was created. The original subscribers came from a list of Maslow's contacts. In 1964, after a successful conference, third-force psychology began nudging its way into the mainstream. It thrived because it brought dignity back to the study of the human condition, focused on self-expression and creativity and in doing so spoke a language that was more appealing than that of stimulus and response, of rewards and punishments.

Humanistic management

In the build-up to humanistic psychology entering into the mainstream, in the 1950s, Maslow busied himself teaching, writing and organising his thoughts. One of his earlier works was the 1954 book *Motivation and Personality*. It was in this book that Maslow argued that individuals would excel in an environment that fostered their physical and mental health (Maslow 1997).

There was nothing remarkable about a psychologist writing a book. This one, a decade before humanistic psychology broke through, was unlikely to make waves. But a funny thing happened. Somebody whom this book wasn't aimed at read it. That somebody was Andrew Kay. Kay was not an academic. He was instead the founder and CEO of a technology company called Non-Linear Systems (NLS). Kay took what Maslow wrote to heart and applied humanistic psychology to his company (Maslow 1998).

First of all, Andrew Kay let his teams decorate their own workrooms. This level of autonomy was frowned upon in the Noughties when my colleagues and I reorganised our desks and computers. Imagine how radical it was in the 1950s when all it took for Robert Noyce to achieve rebel status was to wear a short-sleeved shirt.

Second, Kay dismantled NLS's assembly lines and rebuilt them so that cross-functional teams could work on products from start to finish. Just like Amazon did with AWS, he obliterated the distance, in both space and time, between NLS's creators and their creations.

Third, Kay introduced self-management. Bureaucracy and hierarchy were replaced with a product vision (score). Different but related work was coordinated (orchestrated) by technical leaders as opposed to managers. Interactions and cooperation were encouraged. This wasn't to say that processes and procedures weren't valued. It's just to say that they weren't as important as encouraging interactions between those doing the work.

Kay's focus on autonomy, responsibility and removing barriers to work and how it flowed foreshadowed Hastings and McCord's system of management at Netflix by about 40 years. It was the beginning of humanistic management.

Eupsychian Management

In the summer of 1962, Maslow went to study Non-Linear Systems (NLS). Because NLS had a bank of typists who were happy to transcribe his notes, Maslow produced a journal that he published, more or less unedited, as a book called *Eupsychian Management*. In it, Maslow said that a fairly good organisation will improve fairly good people. Through this mechanism, he wrote, whole industries and therefore whole societies could be improved. In other words, good management improves both the work and the people doing it. The management of work was therefore nothing short of a utopian or revolutionary technique (Maslow 1998).

More than 40 years later, *Eupsychian Management* landed in my lap after it was republished as *Maslow on Management* (1998). It was a timely arrival. It helped me to understand what I had observed at work. This was a relief because I was at that very moment doing my own head in trying to figure out what was happening to me and my team.

In the summer of 2004, my colleagues and I were writing programs and reorganising our work so we could build web-scale applications in a cross-functional team. My observation was that this work somehow helped us to grow psychologically. Our psychological growth led to more risk taking, less shame and more humour, all of which was subsequently reflected in the excellence of the systems we were building.

The confidence from our successes subsequently led to further and greater risk taking, which in turn led to greater learning opportunities and further growth. It was exhilarating. Like Patty McCord, I felt as if I was tripping on wacky chemicals and/or was in love. *Maslow on Management* explained what was going on inside my team and inside my head.

Peak experience

Key to understanding Maslow is the idea of 'peak experience'. Peak experience, for many years, belonged to the religious. Maslow took a closer look at those 'peaking' and came to a remarkable conclusion. They were not in the throes of a religious experience. They were instead mentally well and the peak experience was proof of that.

He came to this conclusion after observing that emotionally

mature people tend to have more peak experiences where, as well as feeling deep love, understanding and creativity, they also operate at the peak of their abilities.

When peaking, the sportsperson is hard to beat. The conductor has the orchestra in the palm of their hands. The computer programmer effortlessly interacts with their machine and does so with little consideration of time, hunger, thirst or, as Bill Gates was said to have showed, their personal hygiene. A 'peaker', as Maslow called them, might feel in love or that they were tripping on wacky chemicals, as Patty McCord was that day on the drive to work with Reed Hastings.

Maslow turns motivation on its head

In *Towards a Psychology of Being* (1998), Maslow said theologians, philosophers, hedonists and even Stoics were united in their belief that pleasure is the consequence of getting rid of pesky needs and wants. This is the 'needs as a nuisance' school of thought.

The needs as a nuisance school says that annoying, irritating needs and their accompanying motivation states get together and, for example, conspire to convert the need for energy into a desire for food, which in turn creates the motivation to go hunting or gathering. Once a person's belly is full, however, the need and the accompanying motivation both vanish. The nuisance is done away with, like the police in the posh parts of London do away with the drunks and the psychic troublemakers are exorcised. With their exorcism, harmony is restored.

What Maslow showed was that the needs as a nuisance school only makes sense for low-level needs. Higher-level needs such as the need to create don't call forth psychic troublemakers. Higher-level needs instead gave rise to a pleasurable tension. It's this tension that compels the painter to paint, the writer to write and, much to their mum's annoyance, the computer programmer to program.

Unlimited motivation

It's within Maslow's higher-level needs that the best kept management secret of all time can be found: *the satisfaction of higher-level needs increases motivation.* Shall I say that again?

The satisfaction of higher-level needs doesn't decrease but instead increases motivation. Excitement is heightened rather than lessened. Appetites are intensified. Those satisfying higher-level needs want more time to paint, more time in the garden, more time to explore, more time at the computer. As higher-level needs are gratified, humans don't come to rest but rather they come to activity. The more a person gets, the more they want. This type of wanting will never be satisfied. This gives rise to unlimited motivation (Maslow 1998), the same sort of unlimited motivation that was on display at Menlo Park.

Human–computer symbiosis

Computer programmers have an internal force that draws them to the machine. The artistic need swells up inside them and their compulsion to satisfy it not only defines their daily work but their whole careers and lives. They are no different to artists or writers.

It was this internal force that made the hackers' lives at MIT worth living (Levy 2010). It drew Bill Gates and his friends to Triple-C's computers. It thrust me into a lifelong love affair with computing machines and what can be done with them.

Any barriers that get in between the programmer and their compulsion to create will either pull them out of the peak experience or stop them getting into it in the first place. This explains why mild-mannered professors at MIT in the 1960s were known to smash their teletypes in fits of rage.

The personal computer, like time-sharing computers before it, removed barriers to the promise of peak experience. The machine was therefore not only a gateway to high productivity but to a higher state of consciousness, too.

Bezos's triumph

Maslow's triumph was to show that the satisfaction of some needs leads to increases in motivation. This is how he turned the needs as a nuisance school of thought on its head. By unravelling this mystery, Maslow planted the seeds of modern management techniques and offered an explanation as to why, by removing the IT infrastructure bottleneck, Amazon tapped into the creativity and unlimited motivation of its programmers. The cloud helped them to satisfy their higher-level needs and that in turn would bring them not to rest but to further activity. The cloud, viewed through the lens of Maslow, is therefore not only a tool of experimentation but a tool of motivation, too.

AWS was simultaneously a sandbox, the world's best IT infrastructure, a gateway to higher productivity and a higher consciousness and, in the scramble for real estate on the digital frontier, a source of remarkable profit. Amazon showed that what was good for its computer programmers and users was also good for the company. That was Bezos's triumph.

Divided opinion

The arrival of B F Skinner at the other end of the corridor from Licklider feels like another funny coincidence in the history of computing. It wasn't actually that strange. Licklider was a psychologist and at that time was in his own psychoacoustic lab. It was only later that he got roped into a project for improving the usability of a radar system, which is what brought him closer to his vision for human–computer symbiosis.

As to Skinner himself, he divided opinion then like he does now. He was on the one hand hated. This was partially because the stimulus of attractive postgrad students triggered a lecherous response. He disgusted Louise Licklider (Waldrop 2018).

It was also partly because of the behaviourists' stranglehold on academia. In the 1960s, every psychology department in England

and the US was headed by a disciple of Skinner (Goldman 2013). Maybe worse than all that, though, was that as late as the 1970s, only 25 years after the Nazis had been defeated, he was still banging on about how human values (such as ethics) were getting in the way of the perfection of humanity and his vision of utopia.

On the other hand, Skinner smashed the monopoly of the mystics who, until the behaviourists turned up, held a monopoly on human behaviour and perfectibility. By doing this, Skinner helped to put psychology on the same footing as the other sciences. This in turn paved the way for Maslow, who described himself, partly to avoid never-ending turf wars, as an epi-behaviourist (Maslow 1975).

The removal of barriers to creativity is the central theme of humanistic management. The McDaddy of barriers to web-scale programming was the IT infrastructure bottleneck. Its removal was the central design rationale of AWS, which let Amazon's engineers get on with their work or, as Bezos said, it allowed their alchemists to do their alchemy.

What we are left with, then, is a simple idea. Humanistic management is about removing barriers to work. The creator and their creations must be kept together, in space and time, if businesses are to tap into their workers' creativity and unlimited motivation. One of the quickest hacks to creativity and unlimited motivation is the computer. This is something that managers like Bob Taylor intuitively understood and something that Frank at the Dutch bank did not.

A utopia unseen

Maslow never got to see what came next for humanistic management. In 1970, he was out jogging in Menlo Park. Not in Edison's Menlo Park but the one in California. I want to believe he was dreaming about utopian management and what humans could do with it. He certainly had at least a decade's worth of work ahead of him and the timing was perfect, given what was happening with the development of the microprocessor around the corner at Intel.

On this day, though, his heart didn't beat the way it had the day before. Abraham Maslow died of a heart attack on the streets of California only a few miles from where Andy Grove worked at Intel. This titan of psychology would not live to see Grove change his theories into a goal-setting system called objectives and key results, which cast a long, Maslowian shadow right through Silicon Valley (Grove 1995).

Maslow started a chain of events that made sure his work would not only be remembered but baked into the fabric of modern life. Zeus had once held dominion over lightning. Thomas Edison, in an audacious theft, dispossessed him of it. Similarly, Abraham Maslow stole peak experience back from the priests and mystics who held sway over it for thousands of years and brought it back to where it belonged. He brought it back to humanity. He was 62 when he died.

Where to now?

Maslow helped me to understand the force that compelled me to program. It wasn't put there by my parents, necessity or my surroundings. It was part of me. The need to take that urge as far as I possibly could, a process he called self-actualisation, was awoken in me by a BBC computer and defined (and continues to define) the course of my life.

Self-actualisation was the thread that Maslow tugged on as he unravelled the mystery of motivation and productivity. That mystery was further unravelled in 2012 when Google and its web crawlers brought the work of the other architect of humanistic psychology into the mainstream of technology like Maslow had once been brought there by Kay.

Chapter 16

The secret life of teams

It wasn't entirely true when I said that Abraham Maslow was the architect of humanistic psychology. There was another. Many considered Carl Rogers to be the most influential psychotherapist in history. Freud, that pessimistic central European, couldn't match the optimistic Midwesterner, Rogers (Cohen 1997).

Whereas Maslow cooked up the idea of peak experience, Rogers, who by the time he passed away had self-actualised into a guru, a wannabe philanderer and a drunk, worked out the further conditions that were needed for creativity to flourish. This concept lolloped around for decades, appearing on and off, and sometimes in disguise, in books about management. It then collided with Google. As 2012 dawned, the humanists and the computer wonks were about to cross paths again.

Project Aristotle

That year, Google went on a mission to find out exactly what made its best teams tick. It did this by initiating Project Aristotle, the aim of which was to find out why some of its teams muddled on while others flew (Duhigg 2016). If Google could answer this question, then it might have a chance of reproducing its most successful teams. After all, through the diversity of thought and experience of their members, teams produce better results, innovate quicker and flag concerns earlier. Who wouldn't want more of that?

Project Aristotle brought together their own team of social scientists, psychologists and statisticians (Duhigg 2016). They needed the statisticians because Google's web crawlers, which had given Amazon such a headache a decade earlier, had mutated and were subsequently used to gather mountains of data on its own people.

Dead ends

The Project Aristotle team started by looking at team composition. They found nothing. The company that specialises in and makes a fortune finding patterns in data found no patterns in its own (Duhigg 2016). It wasn't because they were hidden. It was because they didn't exist. Team composition was a dead end, and that meant all the old rules of thumb about teams socialising and having shared interests were dead ends, too. Teams were successful but not because they ate burgers together (like the team at Intel did) or drank pallet loads of Dr Pepper (like the gang at PARC did).

The Project Aristotle team then turned to the unwritten rules of teams. This seemed promising but there were contradictions here, too. One successful team had a norm that was exactly the opposite of another. For example, one might love a good argument but another would be conflict avoidant. Some successful teams might have had dark senses of humour while others were politically correct. Many of these team norms, like composition, weren't driving high performance. They were incidental and, depending on your perspective, nice to have, but they weren't essential to success.

A starting point

This must have been a frustrating time for the Project Aristotle team. They seemed to be getting nowhere. But then two aspects, both hard to name, started to emerge from the gloop.

The first boiled down to the fact that people on good teams more or less spoke as much as each other. The Project Aristotle team called this 'equality in distribution of conversational turn-taking' (Duhigg 2016) – a mouthful, but a start nonetheless.

The second observation could be called empathy. Members on the best teams had a feel for how someone was doing based on their verbal and non-verbal cues. The Project Aristotle team called this 'average social sensitivity' (Duhigg 2016). Project Aristotle was on to something and, luckily for its team, so too was a certain Harvard professor.

Enter Edmondson

In the 1990s, Amy Edmondson was investigating how companies succeeded when things, as they were over at Amazon, moved at breakneck speed. Edmondson focused on how learning from mistakes drove excellence (Edmondson 2018).

While studying teams at the local hospital, Edmondson discovered that there was a statistically significant correlation between a team's effectiveness and the mistakes they made. The problem for Edmondson was that she expected a negative correlation. The better the team, the lower the number of mistakes. But it was a positive correlation. The better the team, the more mistakes it made.

Edmondson then had the idea that maybe the best teams didn't make more mistakes but they reported them more often. Further prodding showed this not only to be true, but also that the members of the best teams talked openly about what went wrong, with the aim of catching mistakes earlier or, in the future, preventing them altogether.

Psychological safety

In his investigations into creativity, Carl Rogers coined the term psychological safety. Without it, creativity wouldn't flourish. Not long after that, psychological safety was studied in a work setting. The presence of psychological safety reduced anxiety when a company passed through a transformation and, as early as the 1960s, was shown to lower defensiveness at work (Edmondson 2018).

Edmondson's sliver of genius was to build on this earlier work with the hypothesis that psychological safety wasn't about the individual but the group. Further investigation showed this to be true: psychological safety was a group-level phenomenon that Edmondson defined as the belief that a person can speak up without fear of humiliation or punishment. In the exact way that Maslow predicted, this led to risk taking and growth through failure, which came not from the failures per se but because teams felt safe to discuss them.

Why did all this matter to Google? It turned out that conversational turn taking and high social sensitivity, the two norms that Project Aristotle had unearthed, were both aspects of psychological safety. Once the Project Aristotle team realised this, their work and data started to make sense. More than anything else, psychological safety was essential to making teams at Google work (Duhigg 2016).

Psychological safety and leadership

Edmondson wasn't content to define psychological safety. She also wanted to know how it emerged and in which contexts it worked best. Further research showed that leaders create the conditions for psychological safety to appear. Additionally, it was suited to environments where teams deal with uncertainty, creativity and failure, and who coordinate their work with other groups (Edmondson 2018). Psychological safety was, in other words, a powerful tool for complex teamwork – the exact sort of teamwork that those developing systems engage in.

Psychological safety, systems and their development

Unsurprisingly, when I first read about psychological safety, my own experiences, like Google's data, also fell into place. The glue that bound my generation of technical leaders and engineers to those from the past, like Edison and Taylor, was psychological safety.

There are a few reasons for this.

1. Systems development

A system is made up of components. The grid is a system with bulbs, wires, meters and generators. But systems also have unpredictable end users too, like those who, as they did with the microchip and computers, found uses for the grid that their creators couldn't foresee.

When it comes to system design, creativity doesn't only appear in the development of individual components. Creativity also appears when components are recombined into something new. That's what Marc Randolph's son was doing that day on the drive to Scotts Valley. He wasn't creating individual components but instead was creatively reorganising existing components into a brand new system.

Systems design is full of uncertainty because a change in one component can radically alter the system's behaviour and further design. That's what happened to the grid when Edison invented the high-resistance light bulb. The change in the filament meant that the wires flowing into it could be thinner. That in turn meant that the system Edison had in mind disintegrated and in a flash of insight was replaced by the system we have today.

Alternatively, a change in a component can leave the system intact but alters its performance so radically that what was previously unfeasible becomes feasible. This is what happened with the invention of the transistor. It allowed for miniaturisation, which led to ballistic missiles and later microchips. Miniaturisation also led to transistorised computers like the TX-0. Vacuum tubes were swapped out and replaced with transistors. This didn't create cascading changes like the high-resistance light bulb did to the grid's design. Instead, it made it possible for the TX-0 to fit inside a smallish room where the model railway hackers, away from prying eyes, could develop Expensive Calculator.

Another characteristic of a system is that its components are connected by a network, structure or often a bit of both. For example, we know the components of the grid are connected by wires and cables. However, when Edison's protege, Samuel Insull, built a financial network of smaller Edison companies (like Edison

Chicago) and linked them under Commonwealth Edison, he invented the holding company (Wills 2019). That legal construct is part of the system, too.

Finally, the boundary of a system is ambiguous. The boundary at PARC could have been drawn around the computer itself, with its screen, network, keyboard, mouse and disks. But, since what was invented at PARC was used for its own development, the boundary could easily have been drawn around the computer and the people who built it, Bob Taylor's system of management, the first users and eventual customers.

In a system, every component is the 'client' of another. Because there's a connection between components, when a system is in development, each component team needs to have an idea of what the other component teams are doing. This is why Bob Taylor said, 'If you're going to use what somebody else is building, you'd better have some sense of what it's like, because it's not going to work very well the first time or the second time or the third time' (Smith & Alexander 1999).

2. Leadership, teamwork and systems development

This is why Bob Taylor stimulated interactions between the teams at PARC. He knew the only way to get the components to eventually play well together was to get the teams building them to play well together. To this end, he set up internal conferences and knowledge-sharing sessions and made sure, imperceptibly to most, that each team knew what the other was doing. He personally squished festering resentments. Bob Taylor, in other words, created a psychologically safe environment that encouraged teams at PARC to collaborate effectively.

3. Putting conflict to good use

One way Taylor fostered psychological safety at PARC was through a conflict resolution process of his own creation. This process converted class 1 disagreements into class 2 disagreements. A class

1 disagreement is what happens when passions are running high. People's egos are embedded in their work so criticisms naturally feel like a personal attack. A class 2 disagreement is about the problem at hand (Smith & Alexander 1999).

Taylor used to get the combatants in a class 1 disagreement to describe the other person's position. This meant that they had to get inside the other person's work. The only way to do that was to get inside their heads and consider the design decisions they had made and why. This naturally led to people asking themselves, 'Knowing what I know now, would I have made a different decision?' This led to each party learning something from each other and in the process, developing mutual respect. Once the problem was known, ie once it was changed into a class 2 disagreement, each individual could contribute to its solution.

In her research, Edmondson found that candour, transparency, learning from mistakes and conflict resolution all appear in psychologically safe teams (Edmondson 2018). Bob Taylor's method for transforming disagreements fostered all four of these aspects of psychological safety.

4. Hiring learning goal oriented people

Crucial to Bob Taylor's ability to create a psychologically safe environment at PARC was the hiring process he created. He was not only looking for the best computer people around – he was also looking for people who could collaborate, and that meant he needed people who were comfortable speaking up and if not quite as comfortable listening, then at least willing to give it a try. Taylor was, in other words, looking for people who were learning goal oriented.

When it comes to hitting goals, psychologists divide us up into those who are *learning goal oriented* and those who are *performance goal oriented*. Learning goal oriented individuals care about developing their knowledge and skills. Performance goal oriented individuals care about demonstrating their ability.

In environments that require creativity to move the work forward, learning goal oriented individuals outperform performance goal oriented ones. The learning goal oriented computer programmer, with books on their desk or lots of tabs open on their web browser, will get to a better solution quicker than a performance goal oriented programmer who attacks a problem with vigour but, if it's a novel problem (and they are all novel), will hit roadblocks that they won't overcome.

Learning goal oriented individuals are different in other important ways, too. For example, they actively solicit feedback to improve themselves and fear failure less, partly because they're used to it. On top of that, negative feedback, no matter how painful, drives their performance.

Remarkably, the higher the risk, the greater the learning goal oriented individual's creativity (Locke & Latham 2013). This explains what happened at Amazon when it invented the cloud. The prize was massive but so too was the creativity that rose to claim it. The same happened at PARC and Menlo Park.

Contrast that to the performance goal oriented people, who judge their own success by comparing themselves to others and who focus on showing how good they are by demonstrating their competence or, if they have nothing to show off, hiding their incompetence.

In other words, the performance goal oriented people speak up when they've done something well and keep quiet when they've mucked up or failed. In the clinical settings that Amy Edmondson studied, the performance goal oriented individual's urge would have been to cover up errors rather than report them.

Bob Taylor's first step to creating a psychologically safe environment for the system he was developing at PARC was to build a system of hiring. Twenty-five years later, Reed Hastings figured out the same thing at Netflix and Jeff Bezos transplanted David Shaw's hiring process to Amazon (Stone 2014).

5. The relationship between structure, bureaucracy and intrinsic motivation

That wasn't the only thing Hastings figured out. Like Taylor before him, he intuited that bureaucracy had to be minimised and hierarchies needed to be flattened if those working at Netflix were going to remain motivated. Hastings had found in practice what social scientists were at that moment proving experimentally.

Intrinsic motivation is best described as an inner-directed fascination with a problem or task. As the goal-setting researchers Locke and Latham discovered, intrinsic motivation is affected by two aspects of work. The first is about control; the second is about information (Locke & Latham 2013).

Controlling aspects of work is about pressuring the individual to achieve a certain outcome. This could be about getting a task done on time or forcing someone to take part in a performance review. These are normal aspects of work but they reduce the autonomy of the individual and therefore have a negative effect on their intrinsic motivation. At Pure Atria, the company he founded before Netflix, Reed Hastings learned this the hard way. When he started Netflix, he was determined to do a better job, saying that it was important not to become 'overly cautious, risk averse and hierarchical' (Gulati et al 2019).

At the airline my team and I were helping in the Noughties, we were tasked, on the one hand, to disrupt the travel industry. We had to build new components and then combine them into a completely new system. We simulated failure to see what would happen when the system eventually went live and hundreds of thousands of users tried to book a flight after a TV commercial aired.

On the other hand, we had to behave according to the rules. We weren't allowed to move furniture; that was facility management's job. We had to attend performance reviews. We were encouraged to socialise at Friday drinks. But, without a doubt, worst of all, was that if we wanted to test our software at some point in the next four to six weeks, *we had to fill in a form*. This traumatised one of the team so much that he is still banging on about it 22 years later. This had a devastating effect on our motivation in exactly the same way it affected mainframe programmers who were forced into the role of acolytes (Levy 2010).

Reliable information had the opposite effect on us. Shared goals or test data stuck on the wall at the airline had a positive effect on our intrinsic motivation. Clear information provides context and that allows decisions to be made *without a chain of command*. This enables autonomy and, when that happens often enough, it leads to what psychologists call 'high self-efficacy'. In layman's terms, high self-efficacy means that people get better at framing challenges as situations to be mastered rather than dangers to be avoided.

6. Learning goal oriented people and intrinsic motivation

As well as dealing better with failure and feedback, learning goal oriented people have higher intrinsic motivation than performance goal oriented ones. Bureaucracy and a lack of information, which is toxic for most, is fatal for those who are learning goal oriented. If, for example, they bumped into bureaucratic control of computers at the start of their careers after having free and unfettered access to them for their whole lives, you might find that they (ie me) are still banging on about it *22 years later*!

7. Systems must fail in development

Finally, failure is essential to successful system design. The problem is that the countless failures that occur during a system's design are hidden in the end result and are soon lost to history. The importance of failure therefore tends to be discounted, downplayed or ignored, especially by performance goal oriented managers.

This is a mistake. Successful systems are the ones built by teams who seek to make their systems fail in development. These teams know failure leads to excellence, which is why, if failures don't occur spontaneously, they will simulate them. This is why Bob Taylor insisted that the team at PARC should use their own tools, why Edison consistently altered the environment through evolutionary prototyping and why, with its simian army, Netflix simulates failures that otherwise wouldn't occur. There is, in short, a direct correlation between the sum total of the failures in a system's development and its eventual success (Wills 2019). Psychological safety is the alchemic ingredient that turns all that failure into lessons learned and highly innovative products and services.

The tapestry of humanistic management

A number of threads have finally been pulled together. The tapestry they weave is clear. A leader like Bob Taylor begins the process of creating a psychologically safe environment through a hiring process that selects those who are learning goal oriented. The arrival of a gaggle of learning goal oriented individuals, with their inbuilt ability to deal with feedback and failure, is the foundation on which complex teamwork is built.

Once through the door, these deeply and intrinsically motivated individuals have to be allowed to do their jobs. Or, as Jeff Bezos once said, these alchemists have to be allowed to do their alchemy. A surefire way to stop them doing that is to wrap them up in bureaucracy and hide information from them. This is why hierarchies must be busted, the information that was hidden within them released and bureaucracy, hierarchy's deformed second cousin, has to be limited or removed altogether, which is what Reed Hastings did.

Finally, once all that is in motion, the leader's day-to-day work must continue. That work is about fostering communication, minimising the effects of the hierarchy, keeping bureaucracy at bay, resolving differences and sniffing and then snuffing out any simmering resentments or conflict.

The secret ingredient

What is the secret ingredient of teams? This question bedevilled social scientists and management consultants for decades. It has now been answered. What started with Carl Rogers was later painstakingly proven by Edmondson and then amplified by Google when it shared the results of Project Aristotle. At that moment, thousands of pennies dropped in thousands of heads. Those who worked in cloud computing and web-scale application development, and who constantly drove fear out of their teams, finally had a sensible explanation and the supporting research for something they practised but had struggled to put into words. Teams whose members spoke up, listened and cared for each other were psychologically safe. That's what made them productive.

Before Project Aristotle, how teams of knowledge workers should be managed in order to build cloud infrastructure and web-scale applications was still open to debate. That debate, for practical purposes, is over. There's still some resistance from risk-averse, performance goal oriented managers who are drawn to large companies. But, because they're allergic to such managers, computer programmers and cloud infrastructure engineers can, will and do vote with their feet. For those who are learning goal oriented, there's no point in working in a place that tramples all over the thing that they live for: a fascination for tinkering with computers, which is the source of their passion, creativity and remarkable work ethic.

End of Part 4

Once the web arrived, it didn't take long for budding entrepreneurs to start dreaming about commercialising it, or for the White House to put Socks, the First Cat, on its website. This was a confusing time. Some businesses used all the wrong technologies. Others had business models that were mail order versions of existing bricks-and-mortar stores. Some wasted all their money on fleets of branded trucks.

The early period led to financial speculation that overinflated the value of web-based businesses. This created the dot-com bubble that started to burst in 2000. Wall Street finally came to its senses with regard to the valuation of the handful of web-based businesses that had survived the carnage.

Two of those businesses were Amazon and Netflix. They escaped by the skins of their teeth and, although they now seem like regular features of modern life, their eventual success, and the commercialisation of the web, was anything but inevitable at the turn of the century.

After the dust settled, those left building web-scale applications came face to face with an enemy long thought dead. It was possible to host a simple website on a couple of computers but to serve the needs of millions of users all at the same time required vast amounts of computers and the infrastructure that bound them together. This IT infrastructure was stored in air-conditioned and secure buildings that were called data centres but were really just the new computer rooms. They were expensive to build and expensive to maintain. Unsurprisingly, priesthoods emerged around them as they had once emerged around mainframes.

This was an infuriating time for those building web-scale

applications, not least because most of us had come of age in a time of unfettered access to computers. This is why, unlike in the 1960s when it took decades for the priests to be slaughtered, the rebellion was swift.

The charge was led by a bookstore owner who realised that the quick turnaround of features meant user satisfaction, revenue and, maybe most importantly of all, intergenerational customer loyalty. The antidote to the IT infrastructure bottleneck and the new priesthood that Bezos and his team cooked up was an automated system of experimentation that allowed Amazon's programmers to test and learn without delay, failing and growing as they went.

Did Bezos know he was taking part in a humanistic revolution that had started with Abraham Maslow, Carl Rogers and Andrew Kay? Did he know that by giving his programmers a tool to experiment with he was dangling the promise of a peak experience in front of them? Did he know it would lead to bigger risks, richer learnings and emotional maturity, all of which would be reflected in the excellence of Amazon's products and services? Maybe.

If that was part of his thinking, it's not in the history books or any of his writings collected in *Invent and Wander* (Bezos 2021). It's improbable, given the breadth of his reading, that Bezos doesn't know who Abraham Maslow was. But back then it didn't matter. He intuited that something was blocking the creative process and therefore the IT infrastructure bottleneck had to be removed in the most pragmatic way possible. AWS was that pragmatic way.

That AWS went on to become a powerful tool for building web-scale applications and connecting the created to the creator and thus becoming a powerful tool for the psychological development of those using it was, I dare say, not on Bezos's mind. At the time Amazon's web store was on fire. They needed to put all their own fires out but as they did, it dawned on Amazon that it could help other businesses put theirs out, too, by helping them with their own undifferentiated heavy lifting.

Not long after this realisation, Netflix was on fire. In 2008, a database failure stopped it shipping DVDs for three days. Back then, this was how Netflix made all its money. It was this failure that brought Netflix to Amazon's door. At that moment, AWS found its most famous customer and Netflix found an infinitely scalable computer store.

Did anybody really believe that it was the cloud alone that was the secret of Netflix's success? They shouldn't have. A computer, no matter how infinitely scalable, can never be the secret to any company's success. Back in 2007, just as Netflix started streaming, the comedian Eddie Izzard told a joke about guns in the United States. Supporters of gun ownership, Izzard said, love to say that guns don't kill people, people kill people. After a pause for comedy timing, Izzard pointed out that sure, people kill people, 'But I think the gun helps.'

The cloud helped Netflix. The cloud, however, can only ever be a weapon of mass creativity in the hands of those who both know how to wield it and have the courage to do so. Netflix was full of those people, but not by accident. Randolph, Hastings and McCord are unequivocally heroes in the world of cloud computing. They will go down in our history. It may have been Bezos who created the cloud, but it was the executive team at Netflix who put their leadership energy into working out how to use it as part of a system of management that drove high performance. Well before the cloud was a twinkle in Bezos's eye, Hastings and his team had already stumbled onto the idea of talent density and a broad range of techniques to squash bureaucracy and hierarchy. Hastings was also determined to apply the lessons he had learned through his management failures at Pure Atria.

It was in this way that Hastings and his team created a culture that balanced autonomy and responsibility. By doing this, like Andrew Kay before them, the management team at Netflix created the conditions that allowed their people, and themselves, to do their best work, which is one of the surest ways to enter into a peak experience. Although I can't know for sure, this is what I believe Patty McCord was experiencing on her drives to work. I was certainly thrown into a peak experience when I worked at that airline.

As Netflix continued its upward trajectory, and as the whole world was buzzing with excitement about what could be done with the cloud, Google explicitly brought humanistic management into the picture. Google was frustrated because it couldn't quite put its finger on what made their best teams tick. This led to it kicking off Project Aristotle. By doing that, it did every computer wonk and those managing them a favour by bringing the outstanding work of Amy Edmondson into the mainstream of technology. Once the results had been released and processed in 2014, the way in which knowledge workers, the sort who develop cloud infrastructure and web-scale applications, should be managed was so coherent it was practically algorithmic.

First, teams had to be psychologically safe. This wasn't just about taking risks during the development process. It was also about trying out new management techniques. Teams that iteratively built their systems of management from an ever-growing number of best practices found that their cloud infrastructure and web-scale applications co-evolved with their systems of management.

In these safe environments, as Bob Taylor had once discovered, teams took more risks and that led to failures and new lessons. Those lessons were then baked into improved products. This made sense since the more failures in the development process, as Edison taught us so well, the more successful the final products and services would be. Psychological safety therefore explained how systems grew in lockstep with the emotional maturity of those building them. Maslow's utopian technique was alive.

Second, management matters. Incompetent people tend to rise into leadership positions for all the wrong reasons. Such leaders don't see themselves as 'conductors' who have to help write and 'conduct' the 'score'. They are, because of their own emotional immaturity, unable to create the conditions for psychological safety to arise and, without that, high performance and peak experience cannot emerge either.

Managers who succeed with the cloud are the ones who foster a culture of curiosity. Like Taylor and Hastings before them, they eschew bureaucracy and hierarchy and focus instead on talent management, developing a 'score' and in doing so, creating the conditions for unlimited motivation and therefore success.

As 2014 dawned, cloud technologies and the management practices for succeeding with them were both mature, the commercialisation of the web started to look inevitable and the dreams I had as a little boy were coming true all around me. A new age had begun.

Themes redux

In Part 4, the computer shifted its shape again and when it did it went and shifted the shape of the things it touched. It was this shapeshifting nature and the transformability of information that made sure the video store was no longer a place on the high street but a destination online.

Amara's law also struck again. Hardly anybody, including me, thought the commercialisation of the web would succeed throughout the Noughties. As late as 2014, Steve Ballmer, then CEO of Microsoft, said that Amazon was not a real company because real companies make money (Rose 2014).

Finally, just like in Part 3 of this book when we learned a lot about computers, in Part 4 we really learned a lot about humanistic management. We learned about unlimited motivation, psychological safety and the role of the leader in creating it. In Part 3, the computer as we know it today was finally clear. After Part 4, management as we know it today is clear, too, and we now have a frame of reference for understanding the escapades of Bob Taylor.

Where to now?

Vaclav Smil calls the 1880s 'magical'. That decade, more than any other, laid the foundations of the world in which we now live. In the subsequent decades, those foundational technologies merged with changes in public health and labour laws. This led to constantly rising standards of living that continued all the way until 1971, when the final improvements were wrung out of Second Industrial Revolution technologies.

Inspired by Smil, I call the period between 1995 and 2010 the magical Noughties. It was during that long decade that the foundations of modern life were laid. Only six years after the magical Noughties ended, however, the dream took a nightmarish turn when the human race woke up to realise they had not only laid the foundations for the next version of life but may very well have laid the foundations for its final version.

What have we gone and done?

Part 5

THE ENCHANTED LOOM

Chapter 17

The end of the world (of work) as we know it

In 2016, the organisers of Scala Days, an annual conference held in Berlin, asked me to give the keynote. Scala is a programming language – quite a good one, actually – but it's only tangentially related to my interests in cloud computing. I asked the organisers what they wanted from me. They wondered if I had anything to say about technology and society at large.

At that moment, I was reading about post-capitalism, a term popularised by Peter Drucker in his 1993 book, *Post-Capitalist Society*. That's the same book in which Drucker let loose the idea of the orchestra team with its conductors and scores (Drucker 1993).

As I waded through books on post-capitalism, the team at Container Solutions showed it was possible to stitch together powerful open-source tools, such as Docker and Apache Mesos, to automate big chunks of work that used to be done by whole IT infrastructure teams. When Scala Days reached out, I had Drucker in one hand and these open-source tools in the other. Together, they fuelled a little obsession in me about what would happen to society when things that used to have a price became free. As I noodled on all that, something I said and believed could never happen, happened. A computer program beat a grandmaster at the ancient Japanese game of Go.

I found a place to sit down and think. Contrary to my expectations,

I came to the conclusion that, after a stunning run, capitalism was about to come to its inevitable end. I told the team at Scala Days that I wanted to speak about that. They said, 'Great.'

Is it different this time?

I was unprepared for the depth of feeling that the ideas I was playing around with tapped into. I did a dry run of the keynote at a community event in Amsterdam. A man in the front row jumped up with tears in his eyes and poured scorn all over the technologies and people like me who dared to suggest they could do the work of a person. He was an IT infrastructure engineer.

At the speakers' dinner in Berlin, I felt the full wrath of the man sitting opposite me. He called me 'one of them'. He said doomsayers have been peddling the same nonsense for years and machine-gunned me with examples. There was COBOL in the 1960s, fourth-generation languages in the 1980s and object-oriented programming in the 1990s. He said that all of them, and automation in general, were supposed to bring commercial programming (and capitalism) to a sticky end. But none of them had.

Unsure of how to respond, I paused for a moment before saying, 'But this time it's different.' The table erupted into laughter. I breathed a sigh of relief but as I sat there with my sticky toffee pudding, wondering what he meant by 'one of them', I was overcome with dread. I'd just about survived what is normally a sanguine affair, and the next morning I had to face hundreds of people in my first keynote on an emotive topic, my knowledge of which was embryonic at best. What if I was wrong? Even worse: what if I was right?

Reason 1: Jobless recoveries

There were three reasons why I thought I was right. The first was the changing nature of recessions. I've lived through four recessions and their subsequent recoveries. The last four were recoveries unlike any other in history. They were jobless. (See, for example, Jaimovich

& Siu 2012; Graetz & Michaels 2017; or Brynjolfsson & McAfee 2012.) A jobless recovery happens when the gross domestic product – the total value of goods and services a country sells – returns to pre-recession levels, but employment levels do not.

Imagine a single factory. Economic hard times arrive and costs must be cut. The wage bill is the most logical place to start. Later, the economy returns to pre-recession levels and so too do the factory's profits. The jobs make no such return. The factory has learned to do more with less. They have improved their productivity and reduced their payroll. That's a jobless recovery and what happens in one factory happens across entire economic regions.

First they came for routine jobs

Automation traditionally retires routine jobs, which is one way that businesses do more with less. A routine job is one that can be done by following a well-defined set of instructions or what computer programmers call an algorithm.

The automation of routine jobs was once limited to manual work such as the repetitive planting of seeds or the harvesting of crops. However, in the First Industrial Revolution, machines replaced the repetitive although highly skilled weaving and finishing of textiles and, in the last century, driven by powerful computers and improvements in work simplification, robots took over complex but well-defined jobs in factories.

The highly skilled and non-routine work that's carried out in teams, the sort of work that Amy Edmondson said psychological safety lends itself to, has so far resisted automation.

The pace of change and the joblessness of each recovery

Each of the past three recessions sent companies on a desperate scramble to do more with less. In that desperation, companies directed their creativity towards automation, which led to the retirement of not just jobs but whole job categories, which means some roles will never come back.

Before the recovery from the last recession had completed, however, the next one arrived, and so, like a boxer raining down blows, the labour market hadn't pulled itself off the canvas before the next recession arrived and hit it again. This is what happens when technology retires jobs more quickly than the economy can invent new ones.

In 2012, Erik Brynjolfsson and Andrew McAfee said the economy was not putting people back to work. Paul Krugman said that the unemployment situation after the Great Recession was a 'scourge' and a continuing 'tragedy'. Twenty-five months after the Great Recession ended, unemployment in the US was still more than 9 per cent (Brynjolfsson & McAfee 2012). Businesses used new machines but did not bother with new people.

The new alchemists

Should this surprise us? As early as 1982, in *The New Alchemists,* Dirk Hanson described the microprocessor as a 'job killer'. Edison and the alchemists of the previous century married chemistry to physics and in doing so turned science into gold. A new form of society then arrived but less than 100 years later, it gave way to the new alchemists who 'transmuted silicon into digital information' (Hanson 1982; Noble 1979).

Almost 20 years before Jeff Bezos ran around his office screaming that developers were the new alchemists, Hanson's prediction that computer technology would change society as much as the wheel once had looked as if it were coming to pass. This seemed true after the Great Recession, which forced companies to accelerate their own digital transformations. That was the pressure cooker situation that AWS (and its salespeople) stepped into.

The contradiction at the heart of capitalism

Philosophers and economists predicted jobless recoveries back in the 1880s when, in response to the miraculous progress of those years, they wondered where all this change would take humanity.

They came to a stunning conclusion. Capitalism, as a system, will fail if it succeeds (Rifkin 2014).

Capitalism works by pulling everyone and everything into the marketplace. Once there, everything, including people's labour, is given a price so that it can be traded (Rifkin 2014). This was a change for individuals who were historically either free to do what they liked with their labour or coerced as slaves or serfs and therefore had to do whatever their master said.

Once capitalism as we know it arrived, which it did in response to the First Industrial Revolution, workers exchanged their labour for wages, which could be exchanged for food, rental payments and, if there was anything left, the odd luxury item. Those with labour to sell had an incentive to keep their prices as high as possible. Those who bought labour, such as factory owners, had an incentive to keep the cost of labour as low as possible.

One way to keep the cost of labour as low as possible was to maintain a talented and trained but unemployed workforce. For the factory owners, bargaining power was therefore achieved through unemployment, which is a feature, and not a bug, of capitalism.

Factory owners used technology to maintain this surplus workforce. This is why, sooner or later, commercial computer programmers realise that they give weapons to bosses to hit their employees with. I didn't realise that when I helped bring that airline online in the Noughties. I laboured under no such illusion in 2016.

The contradiction at the heart of capitalism is therefore simple to state. The system, in its relentless attempts to do more with less, brings forth technologies such as the transistor, the chip, the computer, and in doing so, it retires jobs. As that process nears its end, recoveries from recessions will become ever more jobless. Once that process has run its course, all work will be automated and all humans will be unemployed. Without wages to buy the goods and services of companies selling them, capitalism grinds to a halt (Rifkin 2014).

A prophecy comes to pass?

Jobless recoveries indicate that a prophecy from 1930 is coming true. That year, the economist John Maynard Keynes wrote an essay, 'Economic Possibilities for our Grandchildren' (Keynes 2009). It was in this essay that Keynes introduced the idea of technological unemployment, which means 'unemployment due to our discovery of means of economising the use of labour outrunning the pace at which we can find new uses for labour'.

This says, in plain English, that humans will invent machines that retire jobs quicker than capitalism can invent new ones to replace them. In other words, workers will one day in the future lose the race against the machines. As the decades passed, economists were left scratching their heads. When will Keynes's moment arrive? The question we have to ask now is, did it just arrive and if it did, what are we going to do about it?

Reason 2: We are living in a dematerialised world (and I am a dematerialised girl)

In the past decade, technology transformed not only the nature of economic recoveries. Technology also changed the nature of products and services. This second reason made me think things were different this time.

A car makes it from the drawing board to your driveway through a laborious, eye-wateringly complex and expensive manufacturing process where it's assembled, tested and finally shipped. All those doing the assembling, testing and shipping earn wages that go back into their local economies. A dematerialised product, like a book moving across the internet to a Kindle, is assembled and shipped at the speed of light. Its delivery costs are limited to just a fraction of a cent's worth of electricity.

Because they don't manufacture or ship physical goods, companies that deal in dematerialised products have small workforces. This concentrates wealth in fewer hands while increasing the number of those made unemployed by technology, such as video shop workers like Ray. Erik Brynjolfsson and Andrew McAfee call this 'the spread and the bounty' (Brynjolfsson & McAfee 2014). Dematerialised products increase the bounty (profit) but decrease the spread of wages for workers, which is how in 2022 Netflix generated a whopping $31 billion in revenue while employing only 12,000 people.

In the olden days, there was an upper limit on the bounty that a factory could capture because there was an upper limit on how many cars or sweets their production lines could spit out. The bounty that factories did capture was spread right across the cities in which they were located. Netflix has no such upper limit. Cloud computing allowed them to scale their revenues and profits without scaling their workforce. The bounty is massive, but the spread is minuscule. If all companies manage to do what Netflix does, then the spread will tend towards zero, recoveries will become more jobless and capitalism will grind closer to its inevitable halt.

Serviceisation

Once products were dematerialised and could be delivered as bits in a digital stream, services and subscriptions became possible. The first business to get this right was Netflix, when it introduced a subscription service and cannily used its customers' living rooms as a warehouse (in the same way that Uber cannily uses its drivers to take care of oil changes, the tyres and cleanliness of the taxi).

Over the years, subscription services became complex but the principle remains the same. For example, Amazon started its Prime service in 2006. Unlike Netflix's streaming service, Prime is a multi-service subscription that provides its customers with the same-day delivery of goods, streaming music and video, ebooks and, of course, shopping services. Amazon's Audible is an additional service for audiobooks.

Once signed up for one service, it's difficult for customers to not sign up for *just one more*. When this happens, businesses further ensnare their consumers into their ecosystem, where they want them to spend as much time (and money) as possible. This is how I went from owning an iPhone to signing up for nearly all of Apple's services. At no point did the extra money I spent appear on somebody's payslip.

Serviceisation, including the delivery of software as a service (SaaS), exacerbates the spread and bounty problem by allowing companies to further divorce their revenues from their headcount. Through that mechanism, more gains accrue to shareholders and other investors as money is sucked out of the pockets of workers. The rich, in other words, get richer. The poor get YouTube Shorts.

Reason 3: Machines can play go

In March 2016, a third reason slapped me in the face harder than Amazon had once slapped the high street. Seemingly overnight, the fever dreams of a million crackpots came true and what once seemed improbable – that a machine could be made to think – was all of a sudden probable.

Top-down artificial intelligence
Artificial intelligence hype was in overdrive in the 1960s and 1970s. There were two main schools of thought. The top-down school focused on 'symbol manipulation', which was about sticking facts into a database and then querying them.

The problem was that, no matter how many facts a computer program had, it was still daft – and when I say daft, I mean *really daft*. A human child can work out the gaps in language. A computer cannot. For example, most English speakers know 'time flies like an arrow' is an expression that means time seems to pass quickly. A computer program might interpret that sentence as if there's something called a 'time fly' that's a bit like a fruit fly. Since fruit flies like to eat fruit, the computer might conclude that time flies like to eat arrows. Ask a computer to draw this for you and it'll draw a fly nibbling on an arrow. This is not artificial intelligence – it's artificial stupidity.

Bottom-up artificial intelligence

The bottom-up school of thought focused on simulating, albeit crudely, what happens in an actual brain. What does happen in an actual brain? After a period of learning, the brain is able to process mountains of analogue information all at the same time. Put a person inside a brain scanner and give them some fried chicken and you'll see their brain's activity explode like a firework display. Replace the chicken with a labradoodle and the scan will show different areas of the brain light up and, as long as the person isn't allergic to or scared of dogs, they will relax as they stroke it.

Brains pull off the trick of the massive parallel processing of analogue information by using networks of neurons. Neurons are a type of nerve cell that are electrically excitable. They're made up of lots of dendrites and one axon. The neuron 'fires' a charge through the axon when the input from the dendrites reaches a certain threshold.

A miracle of both biology and computing, the neuron takes analogue information in, monitors it and, when a threshold is met, sends out an electrical signal. Neurons, in other words, receive analogue signals as input and produce digital signals as output.

In the brain, neurons are organised into networks that feed into each other. The output of one neuron is the input for another. These are called *neural networks*. As the 1960s ended, bottom-up artificial intelligence researchers focused on creating artificial neural networks. They were a promising line of enquiry.

The hatchet job

Then, in 1969, in a squabble over ARPA's research dollars, a man called Marvin Minsky, who called human beings 'meat machines', did a hatchet job on bottom-up AI. In a book he wrote with Seymour Papert called *Perceptrons*, Minsky proved that a perceptron, the name given to an early type of artificial neural network, could not learn how to carry out an exclusive or (XOR), a severe limitation for any computing system. The problem was that it wasn't true and, worse than that, Minsky and Papert knew it not to be true (Grim 2012).

Minsky and Papert's hatchet job booted artificial neural networks into the academic long grass. The man behind the perceptron, Frank Rosenblatt, was broken by this. He died in a boating accident a short time after, in 1971. He was 43.

Backpropagation

In the 1980s, neural networks began to return from the wilderness. A technique called *backpropagation of errors* made artificial neural networks easier to train. As time passed, they became good at pattern recognition, including facial recognition, although they're still bad at telling the difference between fried chicken and labradoodles, Chihuahuas and chocolate chip cookies, sheepdogs and mops.

Breakout

Nolan Bushnell and Ted Dabney started Atari in the 1970s. In 1976, they released an arcade game called *Breakout*. Steve Jobs and Steve Wozniak helped to build it. They didn't work on a computer but instead used a 12-inch circuit board and a few transistors, which they soldered by hand. *Breakout* came a long way in the 37 years that followed (Mason 2019).

Partly because of methods such as backpropagation of errors and partly because of the continuing effects of Moore's law, a company called DeepMind trained an artificial neural network to play *Breakout*. At first, DeepMind's program thrashed like a baby, making senseless moves. Soon though it figured out the rules and scored a few points. Not long after that, DeepMind's neural network was by far the best *Breakout* player on the planet (Clark 2015).

Chess and go

As early as the 1970s, computers could beat humans at chess. Admittedly, these players were, like me, chess idiots. But a start had been made and, 30-odd years later, in 1996, a computer finally beat a grandmaster, Garry Kasparov. Chess is easy for a computer program because the permutations of the board are relatively small. For example, shortly after the first moves in a chess game are made, there are 'only' more than one hundred million possible moves left. That number is hard for humans to get their heads around. For a computer, it's minuscule.

Go is a different beast altogether. A go board is made up of 19 rows by 19 columns. One player has a bagful of black pebbles and the other a bagful of white. If a bunch of these pebbles, or what go players call stones, surround another colour, then the enclosed pebbles change colour. For example, a black pebble surrounded by white pebbles will become white. The moment before that happens, when the stone is under threat, is called *atari*. The game is eventually won by the player with the most territory.

The number of legal moves in Go is 2.1×10^{170}. That's 21 with 169 zeros following it. The number of atoms in the known universe is a number with 80 zeros following it. After three moves in go, there are still 200 quadrillion possible configurations of the board (Kohs 2017; Suleyman & Bhaskar 2023). I knew that a go program would never beat a human using brute force like a chess program had beaten Kasparov. That would not happen in my lifetime and most probably not in anybody's lifetime. I also knew that go would never be beaten by a computer because I 'knew' top-down AI was a dead end and that there would never be enough computing power to train artificial neural networks. I was right about a method using brute force and top-down AI. I was wrong about artificial neural networks.

AlphaGo vs Lee Sedol

After DeepMind created their *Breakout*-playing artificial intelligence, they created AlphaGo. DeepMind's researchers trained AlphaGo on a number of human games. Once it became an OK player, like the chess-playing programs of the 1970s, the team at DeepMind made copies of it and got it to play against itself. Because it's a computer program, AlphaGo experimented with millions of new games and did so quickly. Every time it did some mad, random move that worked, it remembered. Every time it did a mad, random move that didn't work, it remembered not to do that again.

When it finally came up against one of the world's best go players, Lee Sedol, it made what looked like mad, random moves. But they were neither mad nor random. In the first game against Sedol, AlphaGo made a move that looked as if it was a mistake but as the game came to its end, it paid off. In the second game, AlphaGo made its famous move 37. The commentators said move 37 was 'unthinkable' and 'very surprising' (Kohs 2017). Sedol returned from a cigarette break. The game stressed him out, which is where the program had a distinct advantage over him, as it doesn't have emotions. As he sat back down and saw move 37, Sedol smiled wryly. He then took 12 minutes to make his countermove. It did not help. Move 37, a genuinely insightful move that go players now practise regularly but no human had ever made before, won AlphaGo the game. AlphaGo went on to win four of the five games in the series it played against Sedol as millions of people watched live on TV.

It's different this time

Right now, capitalism cannot invent enough jobs to make up for the ones technologies have retired. Because technological innovation has sped up, workers are 'losing the race against the machine' (Brynjolfsson & McAfee 2012). There's no reason to think the process of people losing races to machines, like Sedol did, won't accelerate; in the same way that the team at PARC used the tools they invented to create the next versions of themselves, artificial intelligence will accelerate its own adoption and further development.

On top of this, because of its pattern recognition abilities, artificial intelligence will not only retire routine jobs, like the car-building robots of the last century, but non-routine jobs, too. Eventually, the final bastion of human endeavour, complex and creative work, will fall. On 14 November 2024, Brian Porter and Edouard Machery released a paper whose title said it all: 'AI-generated poetry is indistinguishable from human-written poetry and is rated more favourably'.

In 2016, the world turned on the axle of a machine's ingenuity. It will never, ever be turned back. Things are not only different this time but will never be the same again.

Where to now?

Because they don't swim in a soup of slow-moving neurotransmitters, the speed at which artificial neural networks can train themselves is remarkable. DeepMind showed us that. Their computer program is not only a good *Breakout* player – it's by far the best player in the world.

The history of computing teaches us that it doesn't take long for a general-purpose technology like the microchip, computer or internet to be widely adopted. Because they were parts of systems with unpredictable users, those technologies also taught us that it won't be the creators of artificial neural networks that dictate their future, but their users. We can't predict what they'll come up with. But like Amazon and Netflix once did, they'll come up with something, and, if history teaches us anything, that something will turn the world upside down.

How does a business prepare for this future that it cannot predict? That question was already answered during the time of Reaganomics and the collapse of American industry. It is to that answer that we now turn as we draw lessons from the past that might just help us with our future.

Chapter 18

Only the paranoid will survive the coming wave

In 2023, well after all the engineers at Container Solutions discovered and dabbled with ChatGPT, I thought I'd better have a go myself. I was doing another keynote, only this time at a conference that Container Solutions organised. I was at that moment in the middle of the research I needed for this book. My schedule said that by then I should have made it all the way back to 2012, but I was stuck in 1906. My awful understanding of electricity and how electrons flow across a vacuum, a topic that's beyond the capabilities of the organic neural network that sits in between my ears, had me going around in circles.

I needed a break from Lee de Forest and his triode and so logged into ChatGPT. In preparation for the keynote, I wrote some Java. It had been a while. I worked with ChatGPT like I used to work with junior programmers. I started with a few ground rules for the design of the code. It said it understood the rules. Not long after, when I saw its code, I realised it did not. I told it off. Then I swore at it. I still feel bad for swearing at it.

Our teacher–student dialogue continued. It didn't take us long to get into full flow. Together we wrote a fully tested system complete with a graphical user interface, something called 'build scripts' and a web front end. The system monitored rugby results and kept a league table up to date.

Doing this alone, and doing it properly, might have taken me a couple of weeks, maybe even three or four. Some of that time would have been spent on re-learning Java's libraries. A considerable amount would have gone on finding and verifying test data. ChatGPT did that bit in seconds.

No matter how I looked at this experience, ChatGPT and I had taken 20 minutes to do what would have taken a capable programmer hours, days but more likely weeks to complete. Artificial intelligence had gone from playing go to solving real business problems like computer programming in a period of time so short that nobody seemed to have noticed. It felt like the perfect time to panic. I couldn't understand why nobody else had.

The next wave

Actually, at least one person had. Mustafa Suleyman, the co-founder of DeepMind, was, in 2023, a worried man. He wrote a book that year called *The Coming Wave*. Waves are appealing as a metaphor for technological change. The wave of electricity swept away centuries of tradition. The transistor wave brought with it radios, integrated circuits and computing machines. When it receded, it left the pristine personal computer on the shores of humankind. In the foothills, however, the mainframes were smashed to pieces and their priests all drowned. But the discrete nature of technological waves is an illusion.

Computer technologies are used to build the next versions of themselves. Each wave contains the power of every wave that came before it and adds some extra of its own. These waves are thus more comparable to a supercharged tsunami whose latest wave is, by definition, the most powerful, but not as powerful as the one that will come next. Suleyman's title downplays the severity of the contents of his book.

The latest wave of this tsunami, which for most of the last century was thought impossible, is artificial intelligence. If we can crack intelligence, the team at DeepMind thinks, then intelligence will crack everything else (Suleyman & Bhaskar 2023). In 2016, I predicted this would cause more upheaval than most businesses could cope with. By the time I spoke again in 2023, after my experiences with ChatGPT, that prediction had come to pass.

Is there any way that businesses can survive?

Since 2014, I have helped companies to prepare for and succeed with cloud computing. In 2022, my clients and friends started to slowly ask me about artificial intelligence. By 2024, it was at the forefront of their minds, not least because the economic crisis of late 2022 and 2023 demanded (again) that they do more with less. One way they thought they could do that was with artificial intelligence.

Doing more with less was not the only thing on their minds, however. History was repeating itself, only this time many of those who run businesses remembered the upheaval as the waves of the web and the cloud crashed around us. Once again, companies worried that they'd be Amazoned, worried that their competitors would get ahead in the artificial intelligence race or, worse of all, worried that a digital upstart would rise up and do to them what Netflix once did to Blockbuster.

I found myself telling a rebirth story that in the world of computing has taken on mythical proportions. Like all myths, it contains a stark warning: its protagonists are in denial of, and later humbled by, a power they cannot fathom, and finally the old is washed away so that the new can be born. Here is that story.

Computing's flood myth

Robert Noyce and Gordon Moore built Intel around Moore's 1965 observation that the number of transistors on a chip doubled every 18 months. Soon after, a consensus emerged that this trend would continue indefinitely. By the time 1970 rolled around, it would be possible to manufacture large-scale integrated (LSI) circuits with more than a thousand components on them.

The bet

Robert Noyce and Gordon Moore thought the latest wave of the tsunami was building on the horizon so they laid a bet. They bet on large-scale integrated circuits, they bet that semiconductors would soon be able to do things that were previously off limits and they bet that one of those things would be computer memory.

Noyce and Moore each invested $245,000 in Intel. Arthur Rock, the investment banker who had previously helped set up Fairchild, invested $10,000. On the back of Noyce's reputation, Rock easily raised another $1.5 million from private investors (Berlin 2005). Andy Grove got a salary and stock options (Tedlow 2006).

'Made in Japan'

As we know, that bet was soon shown to be a winner. By the late 1970s, however, the Japanese were no longer happy to be transistor salesmen (Miller 2022). They wanted to produce semiconductors, too. The old arrangement, which saw the Americans design and manufacture chips that Japanese companies like Sony built into creative products, was therefore coming to an end.

Intel thought this was a joke. Like many in America, they didn't see Japan's interest in computer memory as a credible threat. They instead dismissed it, like everything else 'made in Japan', as shoddy (Berlin 2005). This exceptionalism, this lack of paranoia, would cost them dearly. It was, however, not the only thing blinding them.

The success of the 8086

In the semiconductor industry, a design win occurs when a salesperson convinces one of their customers to design a product around one of their components. These products were things like farm irrigation systems, automated counting systems like Bill Gates and Paul Allen's Traf-O-Data or desktop calculators, the exact sort that Intel had helped Busicom with in 1970.

In 1979, design wins for Intel's 8086 microprocessor, which had evolved from the original they designed for Busicom, were few and far between. Intel was getting a kicking from both Motorola, whose chip was better, and their own customers, who seemed to delight in telling them that Motorola's chip was better (Jackson 1997).

What was not better was Motorola's customer service. Intel thought they could beat Motorola here so they conceived a sales and marketing initiative called Operation CRUSH. They would remind the market of just how good their service was and, while they were at it, double the number of design wins for the 8086. CRUSH resulted in a remarkable 2,500 design wins. None were more significant than Earl Whetstone's.

Mike Markkula's Mondays

Not long after soldering together Bushell's *Breakout*, in 1976, the two Steves launched Apple Computer. Mike Markkula assisted them on Mondays. Markkula enjoyed playing tennis, building furniture, having another baby and, on Mondays, providing free advice to entrepreneurs. He could afford this because, when he retired one year earlier aged 33, his Intel stock was worth more than $10 million in 2024 money (Berlin 2017).

After meeting the Steves, Markkula contacted Bob Noyce. Markkula wanted to present Apple's computer to Intel's board. They let him, but after getting burned with a disastrous attempt to sell Microma watches, Intel had no appetite for getting involved with another consumer product. The computer impressed one board member, Arthur Rock, who thought he saw a sliver of potential. He later invested $60,000 in Apple.

Ann Bowers' foresight

Rock and Markkula were not the only links between Intel and Apple. As 1976 turned into 1977, Ann Bowers set up her consultancy business. Bowers left Intel after marrying Noyce because she didn't think it was appropriate to be married to the boss.

Bowers agreed to consult to Apple and bought some of Wozniak's stock. Bob thought Ann was nuts. He had an inkling that personal computers might one day be big, but there was no way these two hippies, aged 21 and 26, would be the ones to make them big. Xerox, maybe, but these two? No way. At the time, Wozniak, the phone phreaker, stole telephone calls using a device that emitted electronic pulses; and Jobs had an unkempt beard, long hair and wore jeans and Birkenstocks. Noyce, Silicon Valley's original rebel, the mould from which all others were cast, who once had the audacity to wear short-sleeved shirts to work, baulked at these two who didn't even cut their hair short. Ann, of course, saw past the sandals and long hair and in doing so saw something that the great predictor and shaper of the future, Robert Noyce, did not (Berlin 2005).

The most far-reaching decision in the history of computing

By 1980, Apple's success proved Ann Bowers, Arthur Rock and Mike Markkula right. They had seen a wave forming and, in Apple, had backed and helped shape a winner. The speed with which this happened caught them all off guard. It was this epic and surprising rise that, a few years later during Operation CRUSH, opened the door for Earl Whetstone. Whetstone was a sales engineer. Like many at Intel, he was comfortable taking risks and with Grove in charge, was encouraged to do so. This is exactly what he did when he asked for, and unbelievably got, a meeting with IBM, the most successful computer company in the world.

At that time, IBM considered asking an outside vendor to build a microprocessor for their upcoming line of personal computers. They had never done this before. Everything they had previously built and sold was proprietary and therefore impossible to clone. In 1980, though, they were in the embarrassing position of playing catch-up to that Californian pipsqueak.

Following the same logic, IBM also asked a software developer that specialised in writing code for microprocessors to write the operating system. The resulting personal computer brought together the hardware company, Intel, into a long-running association with what was then a tiny software development house called Microsoft. Unbeknown to all parties, the future architecture of personal computers had just been conceived. The chips would come from Intel and the software would come from Microsoft.

IBM, whose offices Earl Whetstone had walked into on that fateful day, never recovered from this disaster. They exchanged the heart and soul of their new machine, neither of which was proprietary, for a quicker time to market (Tedlow 2006). Shortly after, demand for IBM's personal computer exploded. Shortly after that, IBM's competitors cloned their personal computers, undercut them on price, so demand for *their* machines exploded, too. Inside every one of these machines, cloned or original, was a microprocessor from Intel and an operating system from Microsoft.

The wave breaks

The profits from the sales of microprocessors masked the wave of Japanese memory that broke on Intel's shores. As 1983 changed into 1984, Intel built more factories and hired more staff to meet the forecast demand. When the economy cooled, these actions proved premature and worse, impossible to reverse. At the same moment, the whole industry woke up to the fact that Japanese memory was better quality and cheaper. If Intel tried to fight this battle, it would be its last (Tedlow 2006).

Grove's out-of-body experience

In his book, *Only the Paranoid Survive* (1996), Grove recalled shuffling around in shock for a year. Then, in a meeting in 1985, he asked Moore, 'If we got kicked out and the board brought in a new CEO, what do you think he would do?' (Grove 1996; Tedlow 2006). Without much hesitation, Moore said that new management would get Intel out of the memory business. Grove then asked, 'Why shouldn't you and I walk out the door, come back and do it ourselves?'

Grove's biographer, Richard Tedlow, called this conversation a real moment of truth for Intel and a 'cognitive tour de force' by Andy. Why? Grove had managed to do something that hardly any business manager or leader could or would ever do. He had looked at himself, almost as if he was having an out-of-body experience, and tried to judge the situation objectively. This let him frame the situation in a way that nobody at Intel, in the previous full year of bickering, had been able to do. In a flash, Grove could see the future, and after that conversation, Moore could see it too.

Escaping on the lifeboat of microprocessors

It took years, including the final year of 1986 when Intel lost money for the first time since 1970, but Grove eventually took Intel out of the memory business. By 1992, mainly because of Earl Whetstone's design win, Intel was the world's largest semiconductor company. They had left the now-commoditised computer memory business to the Japanese. The US microprocessor industry, unlike the automobile and electronic consumer goods industries, survived Japan's resurgence. This period was not like a wave, Grove said, but rather like a tsunami of change (Grove 1996).

The mistakes and the lessons

There are important, avoidable mistakes and a few lessons from Intel's experience that apply to those dealing with the receding wave of cloud computing and the wave of artificial intelligence that's currently breaking on our shores. Let's start with the mistakes.

Arrogance

The first was arrogance. The dismissal of Japanese computer memory was ludicrous. From as early as 1978, just after he relinquished day-to-day control of Intel, Robert Noyce had warned about Japanese memory (Berlin 2005). Noyce was gifted with foresight but cursed never to be believed. I cannot help but think that Mustafa Suleyman is presently fulfilling the same role.

Wilful ignorance

The second was that they ignored the data. The success of the 8086, and especially Whetstone's win at IBM, made the crisis (and Bob Noyce) easy to ignore. But the data was all around then. Microprocessors and memory were two distinct businesses. Are we really supposed to believe that their numbers were jumbled together or that they didn't understand the tax windfall that came their way in 1982? That's impossible. They weren't ignorant. They were deluded. Andy Grove and his ability to look reality in the face have both reached near-mythical status in Silicon Valley. Yet, in a frank admission in *Only the Paranoid Survive*, the book Grove wrote about his time at Intel, he said that he attacked the data. It's remarkable that Tedlow (in his outstanding biography) didn't latch onto this and explore it further, because Grove's indecision stalled Intel's progress by at least one year.

Denial

Denial is the most predictable of human emotions and Intel and Grove were not immune to it. This is the third lesson. Despite the remarkable success of CRUSH, the full ramifications of which were in the future, Intel was a memory business. They lived, breathed and dreamed of computer memory. They were tooled up in a multi-billion-dollar manufacturing operation to produce memory. Memory proved Noyce and Moore right. Memory propelled Intel to the stock market and made them both rich. But memory would take them no further. It took Intel more than half a decade to face up to this fact. It only became clear after Grove's tour-de-force conversation with Moore. Even then, Andy struggled to say out loud that Intel was getting out of the memory business (Grove 1996).

How many businesses are currently caught in a maelstrom of

denial? How many are busy attacking the data or the technology that could actually save them? Blockbuster attacked the web. They said it was a fad. By the time they understood Netflix, it was too late.

These mistakes crystallise into three lessons.

Lesson 1: Find an outsider's objectivity

Grove said that an outsider's objectivity allows them to see what those inside a company cannot. This is the antidote to denial. New management will not be better, Grove said. They will simply have no emotional connection to the past and will therefore see clearly what has to be done. It's true that Grove was also in denial but he eventually learned to find an outsider's objectivity. This is the first lesson of Intel.

Lesson 2: Managers have to find hidden depths

No leadership course in the world could have prepared Grove for what happened to Intel. His escapology was instead honed in his childhood. After avoiding Nazi occupation for most of the war, Grove's childhood was shattered when Adolf Eichmann arrived to take care of the Jewish problem himself. András István Gróf, as he was then known, was the child of fully assimilated and well-to-do Jewish parents. His mum kept them both alive during Eichmann's killing spree, then got them through the battle of Budapest, one of the bloodiest of the war, only for the Soviets to arrive.

Years later, as a teenager, Grove escaped Hungary. He passed the white cliffs of Dover in the freezing hull of a ship when Brattain and Bardeen were in Stockholm picking up their Nobel Prizes for inventing the transistor. Who could have guessed what Grove would go on to do with their invention?

Now, despite how tough Andy Grove was, until his death he spoke about the events of 1985 and 1986 as if they were fresh in his memory. Arthur Rock said that the vote to exit the memory business was the most 'gut-wrenching' decision he had ever made as a board member.

Like Grove and Moore once did, those leading their companies through the currently breaking wave of artificial intelligence have to go through the revolving door and come back inside ready for a long and traumatic fight. If they can't do that, they'd better stay outside and let someone else come back through and try. Managers don't have to be as tough as Andy Grove, but they do need to be willing to dig deep into their emotional reserves because technological waves break companies and the people in them. The second lesson from Intel is therefore about having or finding the mental resilience for the fight.

Lesson 3: Companies need a lifeboat

Andy Grove was talented and courageous. Tedlow was right about that. He chose a lifeboat and loaded his whole company onto it. Many leaders would not have managed this. Andy's talents, however, would have counted for nothing if Intel had no lifeboat on which to escape. This is the third and most important lesson of Intel and one of the most important lessons in this book.

Designing your lifeboat

Because Intel had kept microprocessors alive, they had a lifeboat in which to escape. Intel, of course, had no idea that they needed that particular lifeboat. They couldn't have predicted that Mike Markkula would leave Intel because Grove didn't like him, and help launch Apple. They couldn't have predicted that IBM would break the only unbreakable rule they had: never outsource the core components of your product to an outside vendor. But by building up a microprocessor capability, they managed to prepare for the future even though they couldn't predict it. It's possible for other companies to do this, too.

Assess the future and choose a move

The first thing a company has to do is to work out how clear the future is. If it's clear enough, then companies can prepare for it with simple 'no-regret' moves (Courtney et al 1997). A no-regret move is something that, no matter what happens in the future, will lead the company to have no regrets. Operation CRUSH is a good example of a no-regret move. What downsides could that initiative possibly have had? Almost all futures would have benefited from the CRUSH team's efforts.

Right now, teams are dabbling in artificial intelligence, such as 'co-pilots' that help developers to be more effective; this is a no-regret move. Managers using artificial intelligence to help summarise objectives, create reports or schedule rosters are also making no-regret moves.

Reduce risks through experimentation

If the future is a bit less certain and there are instead a few alternatives, the company's job is to lower uncertainty through experimentation.

At the end of my street there's an artificially intelligent Aldi. There are hundreds of cameras and no cashiers. Customers check in with a bank card, put whatever they want straight into their bags or pockets, and leave. A few moments later, the value of the goods is taken out of the customer's bank account.

The Aldi shop is an experiment. It clearly costs them something to run but it also means that, if they have to, they will be able to scale up quickly. The lessons they learn from the experiment in Greenwich will, in other words, be used across the company. Aldi's supermarket is an example of the evolutionary prototyping that Edison so wisely used as he scaled up his work from Menlo Park to New York.

Reserving the right to play in the future

If the future really is uncertain, then a company's job is to reserve the right to play in it. This often involves research and development or pre-emptive investments into organisational capabilities. This is what Xerox did with PARC. They thought the future of the office was digital and they invested heavily in their abortive attempt to reserve the right to play in it.

This is what Intel did, without really thinking about it, with its continued investment in microprocessors. They reserved the right to play in the future, and when that future arrived, they exploited it brilliantly.

Google short-circuited building an artificial intelligence capability by acquiring DeepMind. For the same reasons, Microsoft invested in a company called OpenAI. These acquisitions reserved Google and Microsoft's right to play in the future.

Lessons to learn

A truly remarkable chain of events, including Xerox's epic fail with the Alto, created the conditions that launched Intel and Microsoft into the stratosphere. At the time, they were plucky start-ups who saw the latest waves forming on the horizon. They placed their bets and in doing so dealt IBM a near-fatal blow from which it never recovered. Not long after that, though, their own hubris nearly killed Intel. Microsoft, incidentally, had no such near-death experience. Intel's lessons are relevant today.

The first set of lessons is about the most predictable human response to change: denial. Denial almost cost Intel their whole business, just like it later cost Blockbuster theirs. What's to say it won't cost you yours? It would be the height of hubris to assume it will not. Dismiss Suleyman, like Intel dismissed Noyce, at your own peril.

The second series of lessons is around mental toughness and the sort of on-the-fly flexibility that's needed to navigate a crisis. The sort of leader who can do this is comfortable with risk and can create psychologically safe environments, which, as Edmondson has shown, help companies bounce back from crises (Edmondson 2018, 2024). Grove, again, pre-empted her findings when he said that bad companies are destroyed by crises but great companies are improved by them (Doerr 2018). This makes sense, since crises are in many ways massive failures so they are massive learning opportunities that the psychologically safe can exploit.

Therefore the real lesson of Intel, the lesson behind the lesson, is that psychologically safe companies encourage the sort of risk taking on display in Operation CRUSH. Intel's culture helped them then and would again when they had to exit the memory business. The real no-regret move, then, for companies that are worried about the cloud and AI, is to seed the ground with psychological safety so that teams are ready to navigate the future with the flower of experimentation. That's a move most companies should make now.

Previously in this book, I've said that managers who are drawn to large organisations often have an emotional need for certainty and are performance goal oriented. When it comes to artificial intelligence, this gives businesses the same headache they had when the cloud arrived. Those who run the businesses that are going to be disrupted most by the coming wave of artificial intelligence are the least qualified to lead their organisations through it. Those organisations need new leadership.

The third set of lessons is all about preparing for a future that you cannot predict. Companies need to reserve the right to play in the future. Xerox knew that and bet on it. When that future arrived, they were best placed to exploit it. Somehow they blew it. This doesn't change the fact that Xerox's approach was correct. Google's logic was similar. It acquired DeepMind and by doing so reserved the right to play in the future.

However, companies don't have to focus on expensive investments. What the arrival of the web and cloud taught us is that companies can do well with an experimental approach. Aldi is doing that with its artificial intelligence supermarkets. Netflix once did that with its experimental and stepwise march into the future. So too did Amazon. AWS emerged from a series of simple strategic moves, proof that incremental and relentless improvements do work.

Where to now?

When I wrote a program with ChatGPT in just 20 minutes, I realised that the predictions I made in 2016 had come to pass. Is that really true? It's true for Aldi and its employees. It's true for the Ibis hotel at the end of my street, where artificial intelligence answers the phone. It's true for cloud and application developers, many of whom are nowadays more productive because they're using artificially intelligent co-pilots.

Whether we're willing to face the truth like Grove did in 1985 but couldn't in 1984, the wave of artificial intelligence is breaking all around us. There's time to prepare but, as it once did for Blockbuster, time is running out. Humanity has almost finished writing the latest chapter of what can be done with Zeus's thunderbolts. When that chapter is complete, the future will have arrived and the world will never be the same again. It is to that future, and the final chapter of this book, that we must now turn.

Chapter 19

Unintended consequences

This chapter is about the unintended consequences of the work we all do in computing. If upon reading this chapter you feel an overwhelming sense of terror, don't worry, that's a sign that you're still sane.

Biological uses of artificial intelligence

There's a biotechnological equivalent to Moore's law, known as the Carlson curve. The Carlson curve says that DNA sequencing technologies will increase in performance at a rate as fast as Moore's law says that semiconductor technology will improve. This is why the sequencing of the human genome cost $100 million in 2000 and only $1000 in 2024 (National Human Genome Research Institute 2024).

Because of the Carlson curve, gene slicing and dicing technology might soon be available over the counter, maybe even at Amazon. With an artificial intelligence as their assistant, a hobbyist might therefore be able to cook up a pathogen, possibly by accident, which spreads easier than Covid-19 but is deadlier than Ebola. In a matter of months, more than a billion people will die. When he wrote about this in *The Coming Wave*, Mustafa Suleyman showed that he knows both his history and his technology. It's never the creators of a general purpose technology that define its destiny. It's the users.

To understand how this is possible and why Suleyman sounded

worried, we have to talk about DNA. If you thought the neuron was a wonder of computing and biology, wait until you get a load of deoxyribonucleic acid. It's brilliant. DNA's four 'bases' are cytosine (C), guanine (G), adenine (A) and thymine (T). These bases are connected in the famous double-helix structure. Words in genetics are called *codons* and they are exactly three letters long and can therefore be combined into just 64 distinct arrangements. That number is derived from 4 letters in the alphabet, raised to the power of the word length, 3 (= 64). Possible words include GAT, TAC, CAT. Like in English, not all combinations make actual words. The combinations that aren't viable are called non-coding or 'junk' DNA.

After a process involving messenger ribonucleic acid (mRNA), each codon (word) attracts one of only 20 amino acids. The completed amino acid chain breaks off from the mRNA and then, in a process known as protein folding, collapses into a distinct shape that becomes a hair cell, collagen or whatever else it's programmed to be. In other words, a four-letter alphabet that makes only 64 three-letter words (many of which are non-coding) and 20 amino acids, is enough to fully describe and build a human being (or a cabbage).

Because gene editing is possible and the genome is simple, it's possible to invent chains of amino acids. These could reduce cholesterol, cure AIDS, make crops resistant to diseases and tomatoes grow to the size of footballs (Suleyman 2023).

The problem is that, until recently, researchers could not predict how proteins fold. The so-called protein-folding problem stems from, according to one estimate by Cyrus Levinthal, the fact that there are 10×10^{300} different ways a chain of amino acids can fold into a single protein. The universe would end long before a computer could generate all the possible permutations (see, for example, Creighton 1992 for a breakdown of the mathematics involved). Predicting how proteins fold remained an unsolved and seemingly unsolvable problem that held back the sort of leaps in biology we need if we are to clean up landfills with waste-eating enzymes and avoid the inevitable moment when antibiotics stop working, which (spoiler alert) is in the next couple of decades (United Nations 2019).

In 2016, the protein folding problem was solved. The European Bioinformatics Institute has a database that previously held 190,000 proteins. That was about 0.1 per cent of all known proteins. After it had learned how to predict how proteins fold, DeepMind's artificial intelligence, AlphaFold, worked out 200 million and uploaded them to the database at the same time (Suleyman 2023). As I was finishing this book, AlphaFold netted DeepMind's other founder, Demis Hassabis, the Nobel Prize for chemistry.

Because of Carlson's curve and systems like AlphaFold, Suleyman can see new forms of biological uses of artificial intelligence amassing on the horizon. This is what worries him. The vast cost of computers in the olden days stopped amateurs writing malicious software, in the same way that the vast cost of purifying uranium stopped teenagers building nuclear bombs in their sheds like Noyce and Skinner once built gliders. This is no longer true for biotechnology. I found a beginner's kit online that cost $89. It comes with e-coli, petri dishes and a sticker that says 'BioHack the Planet'.

Assisted by science fiction, from the Jewish folk story of the golem all the way through to James Cameron's *Terminator*, the idea of autonomous, killer robots is irresistible. The good news is this is not going to happen any time soon (although it will happen). The bad news is that a 12-year-old might release a pathogen that kills a billion people, most likely by accident (Suleyman 2023).

The double-edged sword of technology

Less apocalyptic but more likely in the short term is the effect of artificial intelligence on society and what it means to work for a living. When it comes to the labour-saving nature of artificial intelligence, we find ourselves in possession of a double-edged sword. Deskilling a job is essential for the integration of humans and machines in a production line. For example, Amazon's robots whiz around their warehouses, easily navigating aisles. They can do this for the same reasons that cars cannot drive themselves on the open roads. Roboticised warehouses are designed for them. Roads in the real world, like the 1,000-year-old road that runs through Greenwich, are not really designed at all but emerged from old dirt tracks. This is why cars are currently not driverless but their robot cousins in warehouses are. What Amazon's robots cannot do is pack goods. Instead they shuffle goods to their final destination in the warehouse, where a dexterous human finishes the job. In the same way that the machines deskilled the job of weaving, opening it up to much lower-paid women and children, deskilling the job of working in a warehouse drives down the cost of labour.

Once a human is left in just one stage of the process, as they are in Amazon's warehouses, that stage becomes the focus of intensive automation efforts. Currently the final packing and sorting requires the dexterity of human hands, eyes and the neural networks into which the optic nerves stream millions of bits of data per second. Once dexterity and vision are solved, however, the human packers will be replaced.

The road to automation is, in other words, paved with the indignity of deskilling and the lowering of wages until a machine arrives. The pain of this is captured clearly on the face of Lee Sodel and the millions who watched his defeat on live TV and in the historical account of those whose 'blood was in the machine' by Merchant (2023).

This is one edge of the sword. But there's another. In 2023, a system called ScanNav FetalCheck was launched. ScanNav FetalCheck is a handheld scanner whose data feeds into an artificially intelligent software system. The system works out the gestation age of the foetus and that can help to evaluate biomarkers. It is so easy to use that its operators can be up and running after a tiny amount of training.

This technology will not put any healthcare professionals out of work in Uganda, where it's being trialled, because there aren't enough health professionals there and, even if there were, the healthcare authorities cannot afford the sort of medical equipment that's used in richer, Western countries.

Tools like ScanNav FetalCheck only work if the job is first

deskilled. In my own lifetime, I've seen improvements in healthcare that included how blood is drawn, how basic illnesses are diagnosed and how, for example, gastric balloons are nowadays swallowed rather than inserted surgically. These once-skilled jobs are now deskilled. They are all ways in which the National Health Service in the UK is learning to do more with less. Nobody thinks this is stripping healthcare professionals of their dignity. They are improvements in a system full of overworked and exhausted professionals.

Our ambivalence

This double-edged sword reflects people's ambivalence towards technological advancements. Uber, a despised company, although not quite as despised as Facebook, deskilled taxi driving (Jackson 2023). In London, moped riders bomb around the city with a map clipped to their windshield. Their drivers are preparing for 'the Knowledge', the world's toughest taxi test (Merchant 2023). Transport for London's website says, 'There are thousands of streets and landmarks within a six-mile radius of Charing Cross. Anyone who wants to drive an iconic London cab must memorise them all: the Knowledge of London' (Transport for London 2024). Uber drivers don't have the Knowledge. Instead they use GPS. They also don't work with dispatchers. That's done by the Uber application.

From the company's perspective, it's the best of all worlds. Uber does not maintain or look after the cars; that's the job of the drivers, who they also don't employ. This in turn means that Uber doesn't worry about them getting sick, their pesky rights as workers or their career development.

Not long after Uber arrived, passengers got used to the convenience, cheapness and fare transparency. I conducted a short and wholly unscientific survey. I asked a group of friends if they felt bad for the old taxi drivers. Most of them shrugged. They remembered being ripped off in new cities and pointed out that that cannot happen with Uber. Some of them said, 'That's life.'

Continued inequality

Job simplification almost always precedes automation. That leads to a loss of dignity, as once proud and skilled workers are either replaced or forced into a simplified version of their own job. Eventually, once technology catches up, the job itself will be automated. This will certainly happen at Uber. Once cars drive themselves, there will be no such thing as an Uber driver. There will only be Ubers. The company spent one billion dollars on their self-driving cars. They can currently drive on streets whose layout is as simple as warehouses but still have no chance of making it through Greenwich.

The process of deskilling and automation continued on and off throughout the last century. A turning point, however, came in 1971, the year that Intel invented the microprocessor. It was around this point in our history that jobs started to be systematically deskilled as machines and software took over more of the work. Since then, gains accruing to the few who own the companies that own the machinery have increased, while the salaries of those who work for a living have stagnated.

Artificial intelligence accelerates this process, creating more chaos in an already chaotic labour market. By 2030, the management consulting firm McKinsey predicts that 30 per cent of all hours currently worked in the United States will be automated. Between now and then, 12 million workers will seek a different job because their current job will no longer exist (Ellingrud et al 2023). This will pressurise the already high-pressure labour markets they are entering and thus hand further bargaining power to business owners.

Over in the UK, the Institute for Public Policy Research (IPPR 2024) said that without interventions from governments, unions and businesses, there will be a 'job apocalypse'. They predict that the first wave of artificial intelligence will lead to 11 per cent of all tasks being automated. In the second wave, which will arrive in the following three to five years, a full 59 per cent might be automated.

Why are these numbers so high?

Artificial intelligence might be able to predict how proteins fold but, because it requires the dexterity of human hands and the remarkable ability of human eyes, robots will not be folding bed sheets any time soon. So, why, then, is the IPPR so worried? It's because most white-collar jobs don't require any dexterity. They're already digitised and algorithmic. Accounting, administration, computer programming and the generation of reports from unstructured data are therefore more likely to be taken over by computers. The IPPR thinks these jobs will vanish in the next six years. How ambivalent will the accountants I spoke to about Uber be when *their* job has gone? Shall I tell them, 'That's life'?

Before they totally vanish, these jobs will be deskilled as white-collar workers are teamed up with artificial intelligences, like I teamed up that day with ChatGPT. However, in the same way that Amazon will replace its packers with robots as soon as it can, businesses will also replace their humans with computer systems as soon as they can.

Class warfare

Changes in the labour market have already caused a remarkable amount of misery. In the UK, some reports have said that more than four million children live in poverty (Brown 2024). That's almost a million more than in 2010. By the time this book went to print, that number was upgraded to 5.2 million, which is 1 in every 3 children (Halliday 2024). In the United States, in 2023, the child poverty rate increased to 13.7 per cent (Koutavas et al 2024).

Against the backdrop of this horror show, after flying to space, Jeff Bezos said that the 'only way that I can see to deploy this much financial resource is by converting my Amazon winnings into space travel' (Clifford 2018).

Millions of children living in poverty, as tone-deaf entrepreneurs hoard winnings that would never be theirs without the publicly funded internet, is one of the biggest injustices of our times. The worst thing about it is that these people paid for their own poverty. It was government taxes, of which the working and middle classes pay the most, that funded ARPA and the internet and all the other foundational research in between.

Throughout history, the greed of entrepreneurs has contributed to class rage. To diminish that rage, governments looked for a third way between socialism and unregulated capitalism (Drucker 1993). For example, after the miraculous 1880s, the authorities in the United States intervened to find this third way. Bell Labs and the public price setting of electricity arose from government intervention and the threat of government intervention. The Food and Drug Administration (FDA), which forced those who produced consumable goods to do so in a way that prioritised public health over their profit, was another intervention.

Over in Europe, municipalities took control of street cars and transportation. This wasn't because those who ran the cities were do-gooders, although some of them might have been. It was because they were trying to avoid civil unrest by giving something back to everyday people.

The rise of the demagogue

When a third way cannot be found, those stripped of both their dignity and money, who have to look in the mirror at themselves moments before they meet the eyes of their impoverished children, turn to anyone who can give voice to their anger. This is where demagogues enter the picture. Demagogues don't invent social grievances, but they do exploit them.

Throughout the 2010s, populist politicians exploited the legitimate grievances of people who were poorer, whose kids had fewer opportunities and who used to have jobs that were a source of self-esteem. Because governments in Europe and the United States failed to find the third way, populists, both left and right wing, stepped into the space created by their failure.

Populists of all political stripes blame society's elites. However, right-wing populists add an extra criticism: that the elite are protecting a badly defined group, which is usually minorities (like the Jews in Hitler's Germany), European bureaucrats (like in the United Kingdom before Brexit), immigrants or usually a conflated mixture of both (Judis 2016). In other words, populism gathers the 'noble' people against an undemocratic elite. The 'bottom' and the 'middle' go after the 'top' with right-wing populists taking aim at those above them and below them. It is the third, 'lower' group that differentiates right and left-wing populism (Judis 2016).

Unsurprisingly, shortly before this book went to print and in line with my predictions, Donald Trump won the presidential election in the US by promising to solve economic problems that the last generations of politicians refused to tackle. Their failure is now complete. Those in society who are least able to defend themselves will pay an awful price.

Rats in a maze

Apple ostensibly sells computers, phones, watches and smart speakers. It makes a lot of money doing this. But, like nearly all technology companies, what Apple cares most about is subscriptions. Its music has been a streaming service for years. So too are photo and data storage and Apple TV. Once a customer has subscribed to a few services, it's so easy to subscribe to *just one more*. The real goal of most technology companies is to envelop their customers in a myriad of smart devices. This is sometimes called ambient computing.

Some researchers describe ambient computing as a guardian angel (Pierce 2022). Who doesn't want one of those? The angel informs the emergency services if their ward is having a heart attack. The angel switches their house into low-power mode when they go out. It sends them uplifting pictures of cats when their serotonin levels indicate they're feeling blue.

My Garmin watch isn't quite a guardian angel. It is a pretty good coach, though. It prompts me when to run and when to rest. It knows me better than I know myself, such is the sophistication of its sensors and algorithms. When I'm out running, I listen to books on Amazon's Audible. Later, when I'm at my desk, advertisements embedded in the newspaper recommend new books to me. If I do buy a couple on Amazon, using One-Click of course, it recommends the sort of books that I love and I often buy them. Some of them will arrive later the same day unless I buy them on Audible, in which case they'll arrive immediately.

Technology companies need us to think they have our best interests at heart. It's naive to think that they do. What they need, more than anything, is our money and our data. Our money keeps them going and our data is essential to train their artificial neural networks. The data we willingly provide to technology companies is how they understand our behaviour. Once behaviour is understood, it can of course be controlled. The myriad of devices and rewards they dole out are not guardian angels. They are elaborate Skinner boxes. We are rats in somebody else's maze.

Behaviour control

The Ibis and the Aldi are not the only things in Greenwich that remind me that the tentacles of artificial intelligence are reaching deeper into our lives. The other morning I walked past the hotel, along the ancient road that autonomous vehicles cannot presently navigate and past the Waterstones before ducking into the Tube. On the wall was a gigantic advert for Spotify, the music streaming service. It said, 'My Spotify knows my moods better than I do.' This is not strange. My Garmin knows my moods better than I do, too. It also helps me change them.

We learned this from Skinner and the behaviourists. No surprises there. Of course, computers can be used to alter behaviour, sometimes for the better, sometimes for the worse. There's that double-edged sword again.

The researcher Nicholas Christakis, who runs the Human

Nature Lab at Yale University, set up a virtual game. Participants collaborated by sharing virtual money with the others. The human urge to reciprocate would allow those who gave in earlier rounds to receive money in the later rounds.

Christakis then threw a few computer programs into the mix that pretended to be human. These mischievous programs didn't share. They were tightwads. In response, the humans stopped sharing, too. Christakis showed that just a handful of computer programs were responsible for changing thousands of generous people into 'selfish jerks' (Christakis 2019).

What is free will in the age of ambient computing and artificial intelligence? I started this book by talking about the Enlightenment values of free will, autonomy and, no matter how antithetical it was at the time, choice. Do we still have those values? Do those building artificial intelligence systems share those values with us? In the 2024 landslide election victory in the UK, did I really vote for the Labour party of my own volition? I'm not so sure. The one thing we programmers learn early is that we program computers, but computers also program us.

Zeus's thunderbolts and propaganda

Oh, you don't believe me? Let's go back to the opening decades of the last century during behaviourism's heyday. At that time, the British government had nailed its propaganda. Using vivid imagery and reductive messages (the same sort of vivid imagery and reductive messages that were used to popularise the cloud), the British government energised the whole country against Germany during the First World War (Wu 2017).

The collision between the emerging field of behaviourism and propaganda reminded us of some anti-Enlightenment but nonetheless human truths. Humans do believe that received opinions are their own (Winterson 2022). Their behaviour can be modified. In the same way that they like their Garmin watches telling them what to do, humans quite like their government doing the same. The Enlightenment thinkers had shown that choice may be the cornerstone of freedom but the wartime propagandists reminded us that engaging in a cause larger than themselves is an urge that few humans can resist.

Not long after, in the 1920s, propaganda collided with a new invention, the radio. Using the same reductive and vivid messages, governments could now manipulate the mood of a whole nation. Two admirers had studied the British work from afar. Adolf Hitler, at the start of his career, produced posters for art exhibitions and products like soap (Wu 2017). Hitler collected these early lessons from advertising in the book he started working on in 1923. In *Mein Kampf*, Hitler wrote that the strong leader must understand the great masses' ideas and feelings (Wu 2017). A political movement's messages must then be aimed at the least intelligent because that would lead to the greatest number of followers.

The second fan of British propaganda was Joseph Goebbels. Seeing what the British had done with the BBC (the same BBC that gave me my first computer), Goebbels had the idea of creating a cost-effective radio through which he could funnel political messages. Goebbels would, in his own words, use the radio as an intermediary between the nation and the government.

Hitler and Goebbels learned the lessons from those who 'dabble in democracy', which they had to learn if the Nazis had any chance of consolidating power once they first won it with the help of the radio (Marsh 2021; Morley 2021).

The result was the cost-effective VE30, whose innards included glowing vacuum tubes, a speaker and a transformer. The VE30 soon appeared in millions of homes, apartment blocks, factories and shared public spaces. In lockstep with electrification, the VE30 spread through the countryside (Meier 2018).

In a political masterstroke, the Nazis with the VE30 gave the 'new poor', whose wealth and chances had vanished, the rich who needed culture and the downtrodden who needed charming 'into forgetting their cares by listening in', access to a new media (Meier 2018; Morley 2021). In return, the Nazis got direct access to their minds.

The final piece of the propaganda puzzle was programming. Goebbels realised that, because they were so boring, politics and economics had to be kept to a bare minimum. The Nazis' consumer-friendly programming was therefore made up of light entertainment, a sprinkling of politics and propaganda and, only every now and then, a big political broadcast (Marsh 2021). Tim Wu said that, drawing inspiration from Western attention merchants, Goebbels discovered that a 'spoonful of sugar helps the medicine go down' (2017).

Goebbels' algorithm for programming seems to be alive and well on my X stream. Sports stories, pictures of cats and the daftest of people having the daftest of accidents are punctuated with nuggets of politics and economics. Less frequently, big things happen. NATO's members meet for a summit. An election happens. The King speaks. What medicine is all the sugar in my X stream helping to go down?

Cambridge Analytica

In the early part of the 2010s, Cambridge Analytica illegally collected the personal data of millions of Facebook users. From that data, psychographics, a portmanteau of psychology and demographics, were used to create targeted political advertisements into the social media streams of millions of people. It was in this way that the light-hearted entertainment of social media, with its funny cats and celebrity gossip, was sprinkled with political messages. These messages nudged enough people to vote for Donald Trump in the US presidential election of 2016. For its role in the Cambridge Analytica scandals, Facebook was fined $5 billion by the Federal Trade Commission. By the time that happened, Donald Trump had already won the election (*The Guardian* 2017, 2019; *The Intercept* 2017).

Propaganda in Skinner's box

Goebbels and Hitler worked out that it was possible, with a touch of light entertainment and the odd political statement, to verbalise the base instincts and secret desires of millions of disenchanted people. This is how they exploited existing grievances and, with the help of the 'absolutely essential' radio, amplified them into a political movement (Morley 2021).

Back then, Goebbels needed his own Ministry for Public Engagement and Propaganda and his dedicated *Funkwarte*, the Radio Guard, to make sure apartment blocks and factories tuned in for the big national moments. Modern propagandists have it much easier. They do the same as the old-school propagandists, only this time the technology infrastructure includes the web, smartphones and artificial intelligence.

The Cambridge Analytica scandal showed that when good, old-fashioned propaganda meets good, old-fashioned privacy invasion, millions of people can be categorised and their behaviour manipulated. In its use of reductive, vivid and repetitive messages, the campaign of Donald Trump was in many ways just like what Goebbels had done. In one vital way, though, it was different.

What we learned from Amazon and its early work on its bookstore, and Netflix and its recommendations engine, was that just a sliver of personal information was enough to create a highly personalised shopping experience. This is why my Amazon.com is nothing like yours. We're both individually targeted with the things that their algorithms, and the data we give them, indicate we're most likely to buy.

It's the personalisation of websites that makes modern propaganda different to Hitler and Goebbels. In *Mein Kampf*, Hitler said that messages must be written for the least intelligent people. That way they would speak to something in everybody. This is no longer necessary. Adverts, as the Cambridge Analytics scandal showed, can be tailored perfectly to individuals who, if you asked them, would insist they were thinking and acting on their own free will. So, let me ask one more time, did I really just vote Labour of my own volition?

Is there any hope?

Fortunately, there is. The first glint of hope can be found in the conversations that are happening. Government agencies may be slow, regulation may be woefully inadequate and because the destiny of technology lies in the hands of its users and not its creators, we rest under a sword that may at any moment fall. Nevertheless, all around the world, local authorities and governments are experimentally looking for a third way between unfettered capitalism and socialism. One avenue of investigation is *universal basic income*, the idea that everybody in society gets a basic salary, even if they don't work.

As those experiments continue, writers continue to tell the story of the First, Second and Third Industrial Revolutions. Brian Merchant does an excellent job in *Blood in the Machine: The Origins of the Rebellion Against Big Tech* (2023). At the same time, academics like Nicholas Christakis and the Human Nature Lab he leads continue the painstaking work of figuring out what it means to be human in the time of artificial intelligence. Suleyman's book is a bestseller, so somebody is listening. I certainly did.

Society and its institutions are better prepared for the changing times than they were at the dawn of the First and Second Industrial Revolutions. However, the same institutions managed to ignore all the warnings after the Great Recession and, despite the sanctions against Facebook, have not managed to slow down the effects of social media. The propagandist never had more powerful tools to exploit society's currently very deep grievances.

The second glint of hope can be found in the limitations of artificial neural networks. Humans are flexible and quick; artificial neural networks are not. A small child can, with just one look and a prompt from their mum, learn what a labradoodle is and more than that, can extrapolate what a dog is from the same single example. Artificial neural networks require extensive training and even then are not as good as humans. The sort of rapid learning and general problem solving that comes naturally to a human being is currently beyond the capabilities of artificial neural networks. This gives us and our institutions a chance to catch up.

The third glint of hope comes from the fact that we cannot, cost effectively, solve vision or dexterity. This means that professions that require the sort of bandwidth the optic nerve has will be off limits. So too will be any job that requires a human touch.

The final glint of hope comes from the idea that artificial intelligence might solve more problems than it creates. Humanity will not only put it to use in the antenatal clinics of Uganda but will use it to help find solutions for climate change, the cost of living crisis and all the bore and chore work that none of us want to do anyway (Winterson 2022).

Conclusion

The storyline of the 1966 series of *Doctor Who*, called 'The War Machines', is about a computer called WOTAN (Will Operating Thought ANalogue). Installed at the Post Office and soon after connected over a network to computers like it all around the world, WOTAN becomes self-aware and realises the only way to control the degenerates of the Swinging 60s with their miniskirts and mainframes was to deploy an army of robotic war machines (Lean 2016).

The computer taking over, which is not much different to the creation taking over that we saw in Shelley's *Frankenstein*, is a common trope in science fiction. It plays to our deepest fears that the Enlightenment values of autonomy and choice will be taken away from us. Such stories are irresistible to science fiction writers and journalists (and frankly all of us). But unfortunately, they're not helpful right now.

Advances in biotechnology will collide with advances in artificial intelligence. When that happens, novel pathogens will only be one step away. Suleyman is right about this. The storylines of *Contagion* and *World War Z* are therefore much more likely to come true than that of *The Terminator*.

Much more likely than killer robots, and much more pressing, is the impact of technological unemployment on society. Since the beginning of the 1970s, owners have taken home more profits and left employees with less in their pay packages (Gordon 2016). This trend will accelerate as AI's tentacles take further root in society. Artificial intelligence will cause further injustice and chaos, which in turn will increase the grievances within society that demagogues exploit. AI has a dual role here: it's simultaneously the tool of the capitalist and the weapon of the demagogue.

What we saw with the Cambridge Analytica scandal will seem like child's play to what comes next. Artificial intelligence will be used to create both better psychographic categories and highly targeted adverts, some including deepfakes, that are designed to cause outrage in their recipients. This manipulation is straight out of the radical behaviourist's playbook. Skinner has returned to haunt the human race, only this time the box is so elaborate that we don't even know we are in it.

The future seems grim, but there is some hope. First, there are national and international conversations happening. Writers and academics continue the painstaking work of trying to figure out what our future means. Second, radical behaviourism didn't work in the 1970s and therefore it might not work now. Third, technologists like Mustafa Suleyman are reminding those who will listen that it is not the creators of a technology that dictate its destiny but its users, and here I am reminding you, in case you forgot, that one of those users is you and another is me.

END OF PART 5

Three things happened that took humanity, in just ten years, from selling books online to making artificial intelligence available for programmers and the businesses that employ them. The first was the cloud. It carries out computationally heavy tasks, such as serving millions of simultaneous user requests or training artificial neural networks, with ease.

Amazon originally built its cloud capability to help it run its e-store. For similar reasons, Amazon developed, sometimes through acquisitions, artificial intelligence capabilities, which it needed for products like the Alexa smart speaker.

Later, Amazon provided its artificial intelligence capabilities as a service in the same way it had provided computer infrastructure as a service. Through services like Amazon Rekognition (image recognition) and Amazon Polly (text to speech), Amazon democratised artificial intelligence like it once democratised infrastructure. Some people have started to call it AI as a Service (AIaaS).

With AIaaS, businesses don't have to invest in the systems, software and people that are required to build applications that use artificial intelligence. They can instead use their own data and Amazon's services to streamline their HR processes, for example, or get a computer to answer the phone like the Ibis at the end of my street does. Right now it's as straightforward to add natural language processing to a web application as it once was for Marc Randolph's son to add Stripe to his. In other words, the cloud providers did for artificial intelligence what they once did for web-scale IT infrastructure, providing both as services that their customers only pay for when they use them.

Therefore, only moments after the age of AI arrived, the age of really accessible AI arrived, too. It took years for the transistor to find its way into applications that found their way into the hands of the public. That wasn't true for AI. A student in their dorm room now has access to AI systems and models. What miraculous applications might they create? What mischief might they make?

The second thing was the relentless march of Moore's law. Without increasing computing power, the cloud providers would have had nothing to organise as a service and AI might have remained a science fiction. Moore's law meant that by the time the magical Noughties were coming to an end, computers could simulate not a human brain but at least something similar to a frog's brain (Eindhoven University of Technology 2018). Galvani would have been amazed.

Moore's law matters because artificial neural networks, like Volta's batteries, are scalable. More discs created stronger batteries and more computing power makes neural networks more 'intelligent'. The range of uses of artificial neural networks will therefore develop in lockstep with Moore's law.

The third thing was data. Artificial neural networks need to be trained. In this way, they really aren't like brains at all. A child, upon seeing just one dog, of any breed, can from that moment recognise any type of dog. The human brain comes to us pre-designed, like a home bake kit that needs an egg and some water. How was it designed? By billions of years of evolution. Where is this design stored? In DNA, a code that maps bases to amino acids and thus is really not much more sophisticated than how holes in cards map to musical notes in fairground organs.

Artificial neural networks require a laborious training regime and mountains of data. This data was not available until the arrival of the web and the generation of content as either text or images because of cheap digital cameras and networking infrastructure.

These three related breakthroughs are what make the period from 1996 to 2010 'magical'. The foundations for our lives, and artificial life, were laid then.

This places us in an awkward position, burdening us with choices that none of us want to make. Businesses have a lot to worry about but their problems are at least solvable. By preparing their data, experimenting with AI and finding managers who have the guts and brains for the challenge ahead, they at least stand a chance of flourishing in the near future, even though their staff might not. They will become the government's problem.

Speaking of which, like businesses, the government's challenges seem solvable, too. They 'only' have to find a third way between unfettered capitalism and socialism like our great-great-grandparents did. This is, of course, easier said than done. There are politicians and technology entrepreneurs who are radical behaviourists. They don't care about Enlightenment values, they don't care about children living in poverty and they don't care about you.

What should we do? Should we learn new skills before we get replaced? What should we tell our children to do? Should their focus be on creativity, caregiving and developing the sort of general problem-solving skills that computers don't have? This might work for the privileged among us who can afford and have the energy to retrain or coach their children. Most people don't have that privilege.

For now, there's no consensus on the present or the future. Technological unemployment, which has been growing since the early 1970s, is the root cause of the greatest injustice of our times. A handful of people, building on top of publicly funded infrastructure, have amassed vast wealth at the expense of those who work for a living. The same people have helped to build powerful tools of behaviour control. In their hands, we might end up buying stuff we don't need.

In more malicious hands, we might end up voting against our own best interests, demonising and committing genocide against a defenceless minority or marching to war against what we 'know' to be a brute. Since all this has happened before with only the most basic electronic technologies, what on earth makes us think it won't happen again or, even worse, it isn't happening right now? What medicine is all that sugar helping to go down?

I don't know where things will end. But I do know they begin with millions living in poverty while only a handful of companies, with no government oversight, have in their possession the most powerful software ever created. Computing is no longer the stuff of dreams. It's presently the stuff of nightmares.

Themes redux

In this final part of the book, through the troubles of Intel in the late 1970s and 1980s, we learned that humanistic management and especially psychological safety prepares companies for crisis. Since crises are the ultimate failure, they are also the ultimate learning experience. There are many no-regret moves companies can make now to prepare for the future of AI, but creating a culture of psychological safety is the first one to take. Luckily, psychological safety is the means and the end; teams become psychologically safe by practising psychological safety, which it is never too late to start.

The shapeshifting nature of the computer was on display again. Only around 150 years have passed since something in the air was changed by vibrations in a diaphragm to an electrical current. Graham Bell's experiment taught us that information can be transformed into electricity and back again. When that happened, the telegraph's shape was shifted. It became the telephone. These days, computers, even though their architecture is not much different to the personal computers of the 1980s, are powerful enough to simulate artificial neural networks. Charles Sherrington said the networks of our brains are like an enchanted loom where 'millions of flashing shuttles weave a dissolving pattern, always a meaningful pattern, though never an abiding one'. The same can be said of artificial neural networks. Because they're trained on data and then weighted statistically, artificial neural networks can act, as AlphaGo does, in ways that are impossible to predict. The computer hasn't changed shape, per se, but what can be done with it has. That in turn has changed the shape of society. What do those unique human traits that Maslow cared so much about, things like creativity and free will, mean in an age of intelligent machines?

And I know what you're thinking, that's all miles in the future and that future may never come. Which brings us to that other theme, Amara's law. In the 1930s a group of techno-optimists writing for *Fortune* magazine were ridiculed when they predicted, from inside the depths of the depression, a near future of high wages and economic stability. Amazingly, they were proven right and economists, like Keynes, were proven wrong. Those writers showed that, away from attention-grabbing headlines that most of us have learned to ignore, humans are tenacious (Landes 2003). Like they did in the last century, researchers will continue to beat emerging technologies into shape. You remember the original iPhone and how awful it was? It came out 17 years ago. AI is already advanced, is the focus of research by government departments, including ARPA, and is actively developed by the world's best businesses. AI is also in the hands of users, who will dictate its destiny like they dictated the destiny of all general-purpose technologies. In 17 years' time, AI's tentacles, right or wrong, for good or ill, will be further embedded into our lives. History, Amara's law, what we know about the crossover decades and common sense tell us that. We need to get ready now.

Epilogue

In weaving, the 'warp' yarns are held taut in a frame called a loom. The 'woof' thread is then drawn through, alternatively going under and over the warp. When complete, a complex tapestry, like a story, has emerged from the yarns and the simple, repetitive process from which it was woven. Our tapestry is now almost complete.

The automation of weaving marked the beginning of the First Industrial Revolution. The speed by which knowledge was diffused is what made it a revolution and not just another moment in humanity's history. That speed came from the demystification of knowledge.

Mysteries

In England, 'crafts' were once upon a time called 'mysteries'. The skill to carry out a craft was handed down by a master to an apprentice. All were sworn to secrecy. To the uninitiated, therefore, the craft was a mystery. This may seem strange to modern ears but the remnants of these times are evidenced in town centres all over England. 'Guildhall' is a synonym for 'town hall' but the original use of these buildings was for associations of craftsmen, called guilds, to meet in secret. The beginning of the end of the guilds arrived between 1700 and 1750. Those were the years when a new idea emerged and, along with it, a new word.

The revolutions begin

Téchne is Greek for 'skill' or the 'ability to do'. Scholars and gentlemen scorned this form of knowledge. 'Logy', the suffix found in biology, geology and psychology, means 'the study of'. Neurology means the study of strings, picking up the name because scientists like Newton thought vibrations passed down nerves like they do the strings of a violin.

There was no way to study a skill, no way to study *téchne*, outside the master and apprentice system. That changed with the arrival of books and specifically Denis Diderot and Jean d'Alembert's *Encyclopedia*. The encyclopaedia systematically organised crafts so that an amateur could learn, like I learned BASIC, to become a *technologist*.

Once knowledge was democratised, an ever-increasing number of people contributed to the process of exploiting and generating it. That led to new knowledge becoming embedded inside books or machinery, like when what was previously known only to a craftsperson was embedded in the loom or the Spinning Jenny. At the same time, the Enlightenment taught the British and the Dutch to stop squandering their resources and creativity on wars and religious persecution. Once they did that, their creativity was redirected into trade and invention.

The source of our tidal wave

Transformative gadgets, like the windmill that turned up just down the river from my house, and capitalism have coexisted since the dawn of time but they never combusted into global change. Compared to the Industrial Revolutions, earlier moments of technology and prosperity were superficial (Landes 2003). When the systematisation of crafts into technology and an obsession with knowledge dissemination, which placed a public library in my home town of Hull for the first time in 1775, collided with a growing class of capitalists and the Enlightenment ideals of autonomy and personal freedom and altruism, the Industrial Revolution was born (Mokry 2009). The changes were superficial no more.

This was the seismic moment in our history that started the tsunamic process that to this day continues to deliver waves of changes to our shores. It's not a normal tsunami, mind you, but one where each wave is both powered by and more powerful than the last.

This is the context into which the heroes of our story, Thomas Edison, Joseph Carl Robnett Licklider, Robert Taylor, Robert Noyce, Gordon Moore, Andy Grove and Abraham Maslow, were born.

Zeus's thunderbolts

At the turn of the last century, our heroes started to write history's latest chapters. Was their work a continuation of the First Industrial Revolution? Did they simply pick up where the creators of the loom and Spinning Jenny left off? No. Something had changed.

A recurring theme of this book comes from a quote from a 1945 edition of *Scientific Monthly*. Of Robert Oppenheimer and his team at Los Alamos, it said: 'Modern Prometheans have raided Mount Olympus again and have brought back for Man the very thunderbolts of Zeus.'

It's a lovely turn of phrase and a great nod to Mary Shelley. But it was not accurate. Thomas Edison had already raided Mount Olympus and brought back Zeus's thunderbolts and caged them in his grid of restless energy. Before that moment, humans met their energy needs in one of three ways. They burned things, used human or animal muscle or, using water and windmills, converted indirect solar energy into movement.

This completely changed in the 1880s. It was during that decade that a perfect storm of activity forced together the ingenuity of inventors, investment support for their ideas and the relentless pursuit of their commercial use (Smil 2005).

It was this relentless pursuit that changed the telegraph and telephone, and later the vacuum tube and the microchip, from technological gimmicks brimming with potential into useful devices with societal value.

Vaclav Smil said the changes wrought in the 1880s were so staggering that if aliens were observing Earth, they would wonder why a planet that had lain dormant for 4.5 billion years was suddenly lit up and pulsating radiation from right across the electromagnetic spectrum (Smil 2005).

The increases in standards of living that started at that moment lasted all the way until the early 1970s, when the last improvements to those technologies were finally wrung out. We are still living with the economic consequences of those last improvements (Gordon 2016).

The big calculator

Once Zeus's thunderbolts were in the hands of humanity, with Morse's code in mind, it didn't take long to realise that electricity could not only be converted into motion, heat, cold or even beats in a pacemaker; it could also be converted into information. But progress from Lee de Forest's triode to a fully functioning computing machine was slow. The reason was because, in the beginning, the cutting edge of electronics was about information but not digital information. That was boring old First Industrial Revolution stuff. Analogue technology was where the action was at (Dyson 2012). It was, after all, the vacuum tube that amplified signals and thus made the telephone and the radio, and therefore the Third Reich, possible.

Soon after, though, while fighting the same Third Reich, a genuine computing problem arose that only a digital computer could help with: the generation of firing tables. This kicked off an important woof thread in the history of computing that was dragged through the warp threads of our tapestry. That thread was the computer as a big calculator.

Much like Victorian looms, inputs went into these machines and outputs came out. ENIAC and its peers showed just how useful these machines were. Their miniaturisation, however, had to happen if they were ever to be practical.

Human–computer symbiosis

At the same time that ENIAC was being developed, the psychologists stuck their heads around the door and, since they liked what they saw, started to join the party.

Since the very beginning, to avoid the wiping out of the human race by a machine, computers were not allowed to make decisions. This was because nobody thought it was a good idea for badly written code or dicky electronics to accidentally start World War Three. That's why air defence systems were not automatic. They instead worked in symbiosis with human operators, which explains the importance of human–computer interaction: the air defence systems had to give meaningful and unambiguous information to the operator.

This is how Licklider joined the party. As a psychologist, he was more qualified to work on the human side of human–computer interaction than the physicists were. At that moment, an idea took hold of him. Humans and computers could form a symbiotic relationship. In this way, machines would not only make humans stronger but smarter, too. Together, through a trial-and-error process, a process not unlike how users currently interact with ChatGPT, humans and computers would solve problems that neither could solve alone.

Licklider was not the first person to have this crackpot idea. The history of computing is full of them. Licklider was, however, the only person in history tasked with sorting out command and control, because the world's governments were terrified that they were about to kill each other, probably by accident.

That task, as we know, came with a mountain of money. The way that Licklider spent that money would, these days, land him in jail for breaking every procurement law on the statute books. But he was not spending that money now. He was spending it then.

In a remarkable burst of creativity lasting less than two years, Licklider envisioned the future of the computer, envisioned the sort of network that could connect them to each other, battered the 'Intergalactic Computer Network' into a community and while he was at it, tasked and funded them to start building it. From those beginnings, the computer as we know it today was conceived. Its birth then fell to Bob Taylor. Taylor picked up the work on the personal computer where he himself, one year before his sabbatical in Utah, had left off. This time, though, Xerox, and not Uncle Sam, picked up the bill.

By the time Taylor was finished, the virus was well and truly out in the open. Not long after that, Steve Jobs caught it directly. Charles Simonyi spread it to Microsoft. David Shaw caught it in the early 1980s and was determined to do something with it. Jeff Bezos caught it off Shaw and then took it across the country to Seattle when he started Amazon. It incubated in Amazon's cluttered offices until it mutated into a system called Amazon Web Services.

If Licklider had lived just 15 years longer, he would have seen this for himself. He died in 1990 of throat cancer. The old man was 75. Bob Taylor did see it all. He passed away in 2017. We don't know what arguably the most important man in the history of computing made of the cloud. He was 83.

Humanistic management

Humanistic management and its history, even though as important as the history of the computer, is the background to computing's foreground. By the time he set up Menlo Park, Thomas Edison knew how to manage his researchers and equipment makers. He intuitively knew that hiding information would not lead to breakthroughs and the idea of setting up bureaucratic controls never crossed his mind. Why would it?

At Menlo Park, the team grew through their own failures and their ever-growing mental maturity reappeared as the technical excellence of the systems they were building. There's no evidence that

Edison thought much about this. But if Maslow had had a chance to explain it, Edison would have understood in the same way he would have understood psychological safety.

Later, Bell Labs took the management of knowledge workers one step further. Within their highly creative walls, researchers were given freedom, support from their bosses, worked in cross-functional teams and were encouraged to follow their hunches. When the science fiction writer Arthur C Clarke visited Bell Labs, he said that it looked like a large, modern factory. However, it was a 'factory for ideas, and so its production lines are invisible' (Gertner 2012).

Humanistic management took a more concrete turn at Robert Noyce's first company, Fairchild Semiconductor. He took the lessons he learned there with him to Intel. In practice, humanistic management meant the sharing of information and the busting of hierarchy. Noyce killed those birds with a bag of similar-shaped stones. Small meetings with the boss, newsletters, lunch in the cafeteria, which was organised along egalitarian principles (like the car park too), and breakfast and coffee meetings. These meetings smashed hierarchy, spread information and created a culture of honesty, cooperation and high motivation. That culture and the supporting meetings blurred the lines between managers and workers and that in turn fostered psychological safety.

When Noyce eventually started to codify these practices, it became obvious that he was trying to find a balance between freedom and discipline. By doing this, Robert Noyce created the blueprint that would emerge all across the computing industry, including at Netflix, where Reed Hastings sought to balance freedom and discipline through talent density, radical candour and low bureaucracy.

Kay and Drucker

Two other milestones came in quick succession. The first was the moment that Andrew Kay picked up Maslow's book and then built a system of management based on it. Kay was doing in his own way, and at the same time, what Noyce was trying to do over at Fairchild and later Intel. Kay's work, and his bank of typists, led to *Maslow on Management*.

The second milestone came in the late 1970s. Back at Intel, Andy Grove took what he knew about Maslow and Peter Drucker's management by objectives (MBOs) and smashed them together to create objectives and key results (OKRs).

Grove was as interested in performance as he was in helping those at Intel grow, through failure, to higher levels of emotional maturity (Grove 1995). He was refining Noyce's system of humanistic management. By doing that, he was taking serious steps towards implementing Maslow's utopian technique. It all made great business sense, too. Intel's system of management helped with talent acquisition, retention, motivation and personal growth. Then OKRs literally changed the world.

Humanistic management and business performance

At the heart of Operation CRUSH was an OKR whose aim was to win over Intel's sceptical customers. When Earl Whetstone crossed the threshold of IBM that day, the voice in his head may have belonged to Andy Grove but the angel on his shoulder was Abraham Maslow.

Later, when Intel exited the memory business, it did so because of its culture of humanism that was embodied in techniques like OKRs. Unsurprisingly, Edmondson later discovered that psychological safety was a key ingredient to crisis survivability.

This gives Intel a special place in the story of computing. Robert Noyce invented the microchip. He was then in charge at Intel when his team invented the microprocessor. Intel itself was founded on Moore's observation that the number of transistors that could be crammed on a chip would double every 18 months. As if all that wasn't enough, the Intel trinity then created a system of management that foreshadowed what companies, including Netflix, use today.

The trinity ends

Two years ago, I received a newsletter that I never normally read. I don't know why I opened it that day. It was 26 March 2023. A Sunday. It said that Gordon Moore, the last living member of the Intel trinity, had died. He had passed away two days earlier, aged 94. I sighed.

What about Bob? In his autumn years, Robert Noyce agreed to become the CEO of SEMATECH, a consortium that was meant to help American companies improve their manufacturing capability, which, if they were to compete with the Japanese, they had to do.

Noyce worked 60 or 70 hours a week in his attempts to get SEMATECH working in what Margaret O'Mara described as the 'hyperkinetic' Noyce's idea of retirement (2020). He eventually, and against the odds, succeeded (Berlin 2005). In 2011, MIT's *Technology Review* said that SEMATECH slashed the research cost of new chips and was an example of how industry and government can work together (Malone 2014).

In April 1990, he asked the board to find his replacement (Berlin 2005). As Grove had once taken over Intel, SEMATECH now needed a practical, operational leader to replace the visionary Noyce. Once the decision was made, he and Anne, who at that point must have been delighted with her investment in Apple, looked forward to a few months of rest and then their retirement.

A month later, in May, Noyce was back in Silicon Valley giving a speech about SEMATECH. Steve Jobs wanted Bob to meet (and approve of) his fiancée so asked him over for dinner (Malone 2014).

When he returned to SEMATECH, which was headquartered in Austin, Texas, Bob found out that his colleagues had decided that 1 June was going to be Bob Noyce day. They had printed T-shirts with Bob's face on them and underneath the picture the caption read 'Bob Noyce, teen idol'. A photo from the event, surrounded by his adoring colleagues, was the very last one taken of him. Two days later, almost immediately after diving into the pool for his morning swim, the fast-talking and chain-smoking Robert Noyce, the 'mayor' of Silicon Valley, had a massive heart attack and died one hour later (Reid 1985). He was 62.

Noyce should have been awarded the Nobel Prize with Jack Kilby for their invention of the microchip. Unfortunately, it took the Nobel Prize committee until 2000 to recognise the earth-shattering importance of the microchip. Kilby picked the award up alone.

Andy Grove, as always, was in between Noyce and Moore. He passed away in 2016. He was impossible not to admire. When Brattain and Bardeen were being wined and dined in Stockholm in preparation for their Nobel Prize ceremony, Andy Grove was freezing in the hull of a ship as it passed the white cliffs of Dover towards America. He was still a teenager, did not speak English, was a survivor of the holocaust and the bloodiest battle of the war, the battle of Budapest. After that, he had to survive the communist regime, which is how he ended up on that boat; he was escaping before he was 'disappeared' for his innocuous role in a student protest that turned out to be the beginning of the Hungarian Revolution.

Some days later, on a cold day in January 1957, András István Gróf landed in New York. *Time* magazine had just hit the newsstands. On its cover was the Person of the Year for 1956: the Hungarian freedom fighter. Forty years later, *Time* magazine's Person of the Year was the CEO and chairman of Intel, Andrew S Grove. In between, a remarkable transformation had occurred and a remarkable life had been lived.

I loved Andy Grove because he was a fighter. I loved him because he was an engineer who moved to management and inspired me, and thousands like me, to do the same. I loved him because he used goal setting to drop-kick people up Maslow's hierarchy. I loved him because he didn't miss his children's dinner time or bedtimes. I loved him because he was, quite literally, the only man in the world who called out Bob Noyce's BS (Tedlow 2006). I think that's why Bob loved him, too.

Intel is the leitmotif in the history of computing and therefore cloud computing. It was Noyce and Moore who brought semiconductors and the transistor to Silicon Valley. It was Noyce who then bridged the gap between transistors and microprocessors by inventing integrated circuits at the same time as Jack Kilby did. It was Moore who realised that the number of transistors on circuits would double every 18 months. The microprocessor came out of Intel's highly creative walls. Years later, Operation CRUSH, an utterly Groveian idea, brought the same invention to IBM. That one single, fateful decision, to put Intel's chips at the heart of the personal computer, was the exact moment that the work of Bob Noyce, Gordon Moore and Andy Grove finally collided with the work of J C R Licklider and Bob Taylor.

Themes redux

Our story is now at an end. So, let us, for the last time, look at the themes of this book.

- **Technology is overrated in the short term but underrated in the long term (Amara's law).** De Forest had no idea what to do with the vacuum tube. Bell's scientists said the transistor would amount to nothing. In *Fast Forward* (1987), Lardner said that only ten years earlier, the VCR was a plaything, but at the time of writing '30 million of them have taken residence' in the US alone. Noyce shelved the microprocessor multiple times before finally doing something with it. There is now not a human being on the planet who doesn't rely on microprocessors on a daily basis.

  The hype around AI is dying down. Despite the promise, cars don't drive themselves and that partly explains why AI as a technology, at least by the general public, is underrated. It won't be long before the real impact of AI on our society emerges.

- **Humanistic management.** In this book, we have seen that humanistic management is perfect for developing systems that use existing technologies to invent new ones, which is of course what Edison did at Menlo Park and every single company trying to succeed with the cloud and AI is doing right now.

- **Low bureaucracy.** A feature of humanistic management and another theme of this book is low bureaucracy. Bureaucracy is anathema to the sort of creative people who are drawn to systems development and computer programming. Bureaucracy pulls them out of their dreamlike state, which getting lost in their work plunges them into. Those who are learning goal oriented can deal with setbacks, criticism, wrong turns and blind alleys but they cannot deal with bureaucracy. Andrew Kay's engineers, like Edison's and Bezos's, were alchemists and his low-bureaucracy management system was designed to let them do their alchemy in the same way that AWS was designed to let Amazon's do theirs.

- **Flat hierarchies.** Another theme of this book and feature of humanistic management is flat hierarchies. For similar reasons, Kay, like Edison before him, worked out that information tends to get hidden or hoarded in hierarchical organisations. This stops people, especially when developing systems, from seeing the whole. This feature of systems development was dramatised brilliantly in the 2023 film *Oppenheimer*.

- **Strong leadership.** Speaking of Oppenheimer, he saw the whole, too. The campus he set up in the desert of New Mexico was a scaled-up version of what Edison had done at Menlo Park. It was there that Oppenheimer showed he wasn't just a fantastic theoretical physicist but a great manager, too, a word that rarely appears in relation to him but something he unquestionably was.

- **The symphony model.** Around the middle of last century there was definitely something in the air in regard to humanistic management. It was clear that mavericks had to be allowed to do their mavericking, and for that they needed information. One way to provide it was to bust the hierarchy, which Robert Noyce so nobly did at Fairchild and later Intel. What was this model of working called? It took Peter Drucker until 1993 to come up with a name and accompanying metaphor. Drucker said that shared information was a 'score' and the leader functioned as the 'conductor' of an 'orchestra'. He called this the symphony model and in doing so he gave a name to the best way to manage knowledge workers.

- **Psychological safety.** The final piece of the management puzzle fell into place in the Noughties. Edmondson didn't invent the term psychological safety. What she did instead, which almost certainly guarantees her place in the history of computing, is make the astute observation that psychological safety was a group-level phenomenon. Google then amplified her work. At that moment the puzzle was solved: those wanting to succeed with cloud computing and the sort of software development that it lends itself to need to find a leader who can foster psychological safety, bust the hierarchy, remove bureaucracy and conduct an orchestra of learning-oriented 'musicians'.

- **Computers.** Computers are, of course, the leitmotif of this book. I promised myself (possibly foolishly) that I would teach you about computers without teaching you about computers. It was in this way that we learned about compilers, batch processing, debugging and patching. It was how we learned about the differences between analogue and digital signals and not long after that about exclusive or (XOR), too. But we also learned that most computers are not digital. In between our ears sits a remarkable organic neural network that transforms analogue signals from the real world into binary, electric signals that allow us to solve problems, show love, run for the bus and, with barely any training at all, tell the difference between a labradoodle and a bucket of fried chicken.

At the same time, as this very moment, right throughout our bodies, our DNA is unwinding and from a tiny alphabet of just four letters is creating new cells. This process is algorithmic. So, whereas the emerging behaviour of neural networks is hard to predict, the cells dividing in our bodies are following a process barely more sophisticated than the process a fairground organ uses to play music. Does this reduce human beings to computing devices? No. The opposite.

It was never my intention to dismiss human beings, as Marvin Minsky once did, as mere meat machines. Instead I wanted to show that each of us is a whole, feeling, loving, fearing, poetry-writing and roller coaster-riding sentient being. Our consciousness arises from the interplay of our brains and our environments, both of which were infinitely long in the making. My hope was not to convince you that humans are like digital computers but that computing is a wonderful but wholly understandable process that's baked into the fabric of every one of us and the universe itself. Computing is amazing and that means you are amazing, too.

- **Artificial intelligence.** Speaking of machines and humankind, in the last century a lot of management energy went into getting people to act like machines. In this century, we are (finally) trying to get machines to act like people. Humanity will succeed in this. Play down the upcoming wave of artificial intelligence at your own peril.

- **The crossover decades.** Related to Amara's law are the crossover decades. During the crossover decades, weary citizens and distracted politicians fail to make sense of new technologies. In the olden days, that was annoying for Samuel Morse and distressing to the religious leaders of Boston. Currently, the cloud and AI are causing havoc in their crossover decades. The quicker we get through them, the better. What awaits us all on the other side, however, is hard to predict. Less hard to predict is whether we will make it to the other side. We will.

- **Unintended side effects.** Another theme is unintended side effects. There is no better metaphor for the earth-shattering inability of humans to predict side effects than Brendan and I waiting in the rain to get into the video store. The only reason the video store existed was because the film studios chose the wrong price point for cassettes and the only reason there was a cassette was because Akio Morita's Sony invented the VCR. I love the story of Netflix and the VCR because it encompasses the whole of the 20th century's history. It started with the dropping of Oppenheimer's bombs and ended with Netflix, but along the way coughed up the video store, which itself gave us Quentin Tarantino. The story of the VCR is so funny that Netflix should make a show about it. They could call it 'The Story of Sony vs. Universal Studios: Mary Poppins Meets the Boston Strangler', which is what Jessica Litman called her 2006 paper. It's my favourite story in this book because it's a hilarious yarn of the underdog succeeding because the corporate fools repeatedly punch themselves in the face.

    The unintended side effects of AI will be nowhere near as hilarious as the story of the video store. Humanity will not stop AI and its consequences, just like they couldn't stop the insane race to stockpile nuclear weapons.

- **The computer supercharges its own development.** Another central theme and why unintended consequences always go hand in hand with computers is that the computer supercharges its own development. Technology, as the telegraph evolved into the telephone, builds on itself. That's quite intuitive.

Less intuitive is how the computer is used to invent the next version of itself. The design of computer chips improved once computer-aided design software was created. Not unlike the simulator that Paul Allen wrote for Intel's chip that fateful January, computer-aided design software allowed chip designers to play around with their ideas before sending them off to be manufactured. The improved chips were, of course, more powerful than what came before and that in turn led to more powerful computer-aided design software.

What happened to microchips happens to every aspect of hardware and software development. If you factor in Moore's law and remember that artificial neural networks perform better with increased computer power, then there's no reason to think the development of computers and software will not continue to accelerate.

General-purpose technologies and cloudification. Another recurring theme of this book is what happens to society when a general-purpose technology arrives. The last century saw three technologies – the electricity grid, the telephone and the internal combustion engine – arrive at the same time and then collide with improvements in public health. Together, these four things conspired to clear tons of horse manure off the streets, send humans to the moon, give birth to the computer, improve standards of living around the world and almost eradicate infant mortality in the Western world.

The economic effects of those general-purpose technologies finally came to an end in 1970. Humanity has yet to recover and new, general-purpose technologies have yet to have the economic impact that electrification had. Will cloudification be it? Almost certainly not. The cloudification of society will have an impact but it won't come in the form of economic growth.

- **Computers will probably not improve standards of living for most people.** The final theme of this book is that computers not only shift shape but they shift the shape of whatever they come into contact with, and that includes society. Computers are strange machines. Their switches don't actually move (Mason 2019). Transistors are either on or off but there's no lever inside of them. Software controls these transistors and gets them to dance the most remarkable dances. However, this is not much different to punch cards and fairground organs making the most remarkable music. What this means is that, although they are hard to fathom, computers are really just elaborate and programmable machines.

  What we saw in this book, starting at the beginning with Morse and his code, is that computers are brilliant at solving informational problems. Such problems include processing millions of bank transactions, arranging planes into holding patterns, controlling the light signals that get sent down optic fibre cables, controlling traffic lights and simulations, including flight simulations or simulated worlds where knights ride on chickens.

  Because of Moore's law and constantly improving software, computers make information goods cheap. For example, a good software simulator costs millions of dollars to develop but that's still cheaper than building actual planes and crashing them on purpose to see what will happen. Similarly, gene sequencing is a classic information problem.

  Anything informational in nature, or what can be made informational (which is a lot, given how transformable it is), such as a movie or flight simulator, can be copied perfectly without any wear and tear. So although it's expensive to create the first one, like a movie or this book, the cost to develop one more is basically zero. It's this peculiar feature of information goods that makes the lives of everyday citizens more convenient while not necessarily putting extra money into their pockets.

With all that being said, since most goods and services are not purely informational, computers don't hold within them the secret to economic growth. Pet food cannot be transformed into bits in a digital stream and robots cannot fold clothes or bed sheets (yet). This is why the remarkable progress made in computing has not translated into the sort of epic economic growth that we saw after those fateful three months in 1879 when the electric light, the first reliable combustion engine and wireless transmission were all invented.

This is the real kicker. Computers not only shift their own shape but that of everything they touch. Computers are how telephone cables and switches became the internet. Computers then dematerialised the video store. Software and miniaturisation allowed most major electronic breakthroughs, from GPS to the internet, to be shrunk down and put into our pockets on our 'phones'. The cloud and AI have collided with society, where they are dissolving almost every kind of relationship, economic or otherwise, that we take for granted. Where will this end? We don't know.

Afterword

Galvani's frog

Our story began in the workshop of Luigi Galvani. At that moment in history, two strands split from the same thread. The first was neurology. It moved forward through the exertions of scientists like Galvani and his nephew. The second was the study of electricity which, through Volta's work, became front-page news. But the historical joke, never more clear than right now, is that these studies should never have been divorced.

Volta invented the battery. But like Lee de Forest and his triode, he both invented it by accident and did not understand it. The so-called 'weak' charge in the battery was called 'galvanic' because Galvani first discovered it. What flew out of Volta's batteries was therefore a galvanic current that he 'proved' had nothing to do with animals.

The problem was that the wet frog's leg in Galvani's circuit fulfilled the exact same purpose as Volta's brine-soaked felt did in his (Jorgensen 2021). The world's first voltaic pile was therefore the single-cell voltaic pile – the frog's leg was the cell – that Galvani accidentally created. In short, Volta's battery produced a *galvanic* current and Galvani's circuit was a single-cell *voltaic* pile (Jorgensen 2021). You couldn't make this stuff up.

Some 244 years have passed since Galvani began to unravel the mystery of how the body transforms analogue information into electrical and binary signals in the human body. When he did that, he inadvertently gave birth to neurology. In the same 244 years, Volta's batteries evolved into the capturing and later manipulation of electricity.

Computing machines have finally caught up with Galvani. Artificial neural networks are currently about the same size and sophistication as a frog's brain. This explains why they cannot drive cars.

The history of computing, however, tells us that artificial neural networks will not remain only as powerful as a frog's brain for long.

First, computers and the software that runs on them will continue to increase in sophistication. Artificial neural networks are scalable. Like Volta's battery, whose power increased with each additional disc, artificial neural networks scale with increases in computing power. Moore's law inexorably marches on and therefore artificial intelligence inexorably marches on, too.

Second, theoretical breakthroughs, like the backpropagation of errors – which as this book was being finalised won its inventor, Geoffrey Hinton, a Nobel Prize – will continue to arrive as academics and researchers all around the world investigate ways to make artificial neural networks behave more like organic brains.

Finally, as always in the history of computing, computers and their software will supercharge their own creation. The process cannot be stopped. Does that mean we have really awoken to a nightmare? Maybe not. We can be optimistic for two good reasons.

The first is that technological determinism isn't real. Technological determinism is a theory that states humans have no control over what happens to us and our world. No matter what we do, technology marches all over us 'in seven-league boots from one ruthless, revolutionary conquest to another, tearing down old factories and industries, flinging up new processes with terrifying rapidity' (Beard 1927).

Technological determinism is a fantasy. Technology doesn't tear down factories. People do. Robots don't retire jobs. Managers do. Computers don't steal your data and sell it to a foreign power. Only parasites do that. The future isn't determined by technology but by those who set policy, vote in elections, run businesses and work in them. The future is, in other words, determined by us.

The second reason is that today's problems and fears can be a poor guide to the future. As I said earlier, in the interwar years that gang of techno-optimistic journalists writing for *Fortune* magazine said that inventors push the boundaries of the frontier further while entrepreneurs further develop it. Swimming against all economic wisdom of the time (it was 1939, the last year of the Great Depression), they said that luxuries would be commonplace and increases in wages, due to labour shortages, would drive the standard of living for most Americans to previously unseen highs. These predictions were comical. Similar claims made today about the impact of AI seem as comical. I had to listen to the same nonsense about web technologies. I dedicated a decade of my life to them but never really thought they'd work. I agreed with Paul Krugman's comment about the fax machine.

Time proved the *Fortune* journalists right and everybody else wrong. At the time, it was easy to focus on the noise, like the noise we have to put up with about killer robots today, and miss the quiet work that goes on in the background. The *Fortune* articles showed us that the fears of the present can be a poor guide to the future (Landes 2003).

These are two sources of hope. There is one more. The leitmotif of the history of technology is that it's not the creators of a technology that dictates its destiny but its users. You are one of those users. I am another. If we can both remember that, then the future won't belong to the machines, or the frogs, or the technological determinists. It will instead belong to us.

Acknowledgements

My job in this book was to explain how a domino toppled on a workbench in Italy started a chain reaction that ended in artificial intelligence. That is impossible to do without relying, sometimes heavily, on the work of those who came before you. Therefore, before I thank those closest to me, I want to thank those further away, which is to say I'd like to thank the writers and historians who made my job possible.

The tales of Fiddler Dick and John Tarwall appear in chapter 3 of Tom Standage's wonderful book *The Victorian Internet*. Thanks to Standage for digging them up.

The tale of the Lakesiders appears in multiple places but I leaned heavily on Maines and Andrews' excellent treatment of that story in chapter 2 of their book and am grateful for that treatment.

I wrote the first draft of this book completely from memory. Going into this project, I knew there was one known unknown that I'd finally have to face. The vacuum tube. What was it and how did it become the transistor? I put off this investigation into electronics – every programmer's worst nightmare – until it could be put off no longer. Luckily for me, T E Reid and his masterful book, *The Chip*, especially chapter 2, makes the link between the domino that toppled at Edison's Menlo Park facility and how it became the vacuum tube and how that in turn did not exactly become the transistor but how the transistor happened to have all the functionality of the tube without any of its drawbacks. Understanding that the vacuum tube was just a souped-up lightbulb was what I learned from Reid and I am grateful.

There was something else that was unknown to me. When I closed my eyes and wracked my brains I could see the contours of the whole world of computing. It was all there. The Bobs, Noyce and Taylor, PARC, the web, punch cards and looms, disks and their drives. But there was a dark patch in my mind like something was once there but had been erased like one of Johnny Mnemonic's memories. Someone once said to me forgetting someone you once loved is like trying to remember someone you have never met. I could not for the life of me remember J C R Licklider because I had never, by a miracle of probability, bumped into him before or if I had, for reasons I cannot fathom, I forgot about him. I did not know who he was or the chain of events he set in motion. What I came to realise is that there was a J C R Licklider-shaped hole in the puzzle of my mind. Once I started my research in earnest, it was not hard to find him. All I had to do was look to the side of Oppenheimer and Johnny 'JohnNIAC' von Neuman, go back in time from Taylor or forward in time from Vannevar Bush. Like all roads leading to Rome, almost all stories of computing led to Lick. Once I found him, it was only a matter of a tiny amount of time for me to discover Mitch M Waldrop and his stupendous book *The Dream Machine*.

At the risk of offending people in my own community, when I crossed paths with Waldrop I was in that part of the writing process where I was reading, cross-referencing and adding factual oomph to my narrative. This part of the process is even more fun than the showing up and throwing up part. What set *The Dream Machine* apart was that, unlike books about technology written by those of us who work in technology, it seamlessly blended the exposition of the facts with a narrative that constantly returned to its main character, Licklider.

What I learned from Waldrop and Reid, among many others, including Tom '*Bonfire of the Vanities*' Wolfe, proved my hypothesis that the very best books about computers and technology are written by those not in the field. In their attempts to make sense of computers for themselves and their readers, they do a better job of making sense and drawing out the narrative from the facts than those who work in technology can. That's a lesson I will never forget and am grateful for.

Waldrop's *Dream Machine,* from the moment I found it, occupied my desk, alongside Paul Ceruzzi's *A History of Modern Computing.* Both of these books should be required reading in undergraduate computer sciences courses, and my book – and I get the feeling many other books – would not quite be the same without heavily marked-up and dog-eared copies of both.

Speaking of Waldrop and Ceruzzi, Johhny Ryan's name belongs next to theirs, too. I am grateful for Ryan's fantastic *A History of the Internet and the Digital Future.* His chapter 1 pointed me in the direction of Baran and Licklider and got me started with the opening of chapter 6, 'The Wasp and the Fig Tree'. Years later, when I finally got to Berners-Lee and the browser, I picked Ryan's book back up – it still lay on my desk exactly where I had left it – and was unsurprised to find an excellent treatment of the web's birth in chapter 8 which helped me with the end of Part 3.

I am grateful for Brad Stone's *The Everything Store* and *Amazon Unbound.* It's not straightforward to piece the Amazon/Bezos story together. That the Bezoses, Jeff and MacKenzie, did not like *The Everything Store* made me think that Stone had nailed it.

Speaking of which, well after the first proper draft of this book was already finished, I could not stop myself pulling at the threads of ARPA. As I unravelled it and understood it further, I discovered Margaret O'Mara's *The Code.* I had promised Bev, my editor, I would not make any more changes, but O'Mara gave me a tangible link between Bezos and ARPA. His grandad worked there and they spent summers together talking about computers. Once I had O'Mara, whose book pricks the self-aggrandising creation myths of Silicon Valley companies, it did not take me too long to find *Imagineers of War* by Sharon Weinberger, who tells the story of ARPA in a way that most of us in computing have almost certainly not heard. If I had found these books earlier in the process, my book would be better but I am nevertheless grateful that I could upgrade my story based on theirs. I am also grateful that writers like O'Mara and Weinberger are going after the underbelly of our industry. It needs going after, for all our sakes.

I am grateful to Leslie Berlin for her biography of Noyce, *The Man Behind the Microchip*. Like Weinberger, O'Mara, Ceruzzi and Waldrop, Berlin is the real deal. Although I have read many books, if not all books, about Noyce and the gang at Intel, Berlin's is my favourite and the one I trusted the most. What she did for Noyce, Richard S. Tedlow did his very best to do for Grove. Like *The Man Behind the Microchip*, Tedlow's *Andy Grove* was the text I trusted the most and for that I am grateful.

Chapter 18 of my book is called 'Only the paranoid will survive the coming wave'. That is a mashup of Andy Grove's *Only the Paranoid Will Survive* and Mustafa Suleyman's *The Coming Wave*. I am grateful to Suleyman for his treatment of protein folding and the potential of hackers, as the cost of AI systems and bioengineering fall, to plunge the world into a pandemic. A theme of my life and work, and this book, is the tension between what a single programmer can do if they have unfettered access to a computer versus what a whole team in a research lab can do. A single and not necessarily talented hacker could in the near future use an AI system to build systems that a whole lab could not build right now. I am grateful for Suleyman's thinking on this topic and the direction they sent me in.

Jeanette Winterson, in her marvellous *12 Bytes,* makes the link between Frankenstein and Dracula. You cannot escape Shelley's *Frankenstein; or, The Modern Prometheus* if you work in technology. It is dangerously close to hardening into a cliché. Oppenheimer was the American Prometheus. David Landes' classic work was *The Unbound Prometheus*. In Winterson's phenomenal hands, *Frankenstein* becomes a dyad with *Dracula* and together they link the remarkable progress that occurred in the 19th century by spelling out what happened between the release of those literary classics. Eric Hobsbawm called the years 1914 to 1991 the 'short 20th century'. I feel that Winterson has defined the short 19th century and the years it spans are between 1818 and 1897. If I could one day, with decades of practice and a Road to Damascus rearrangement of my neural networks, become a hundredth of the writer Winterson is, I'd die happy. *12 Bytes* is further proof that non-techies are better at writing about technology than we techies are.

Thanks to Amy Edmonson for her pioneering work into psychological safety. She's a legend in my world and I'm not even sure she knows it. In my role as the CEO of Container Solutions I have experienced the phenomenal power of psychological safety and just how hard it is to build and how fragile it is once you have it. In this book I tried to link her work to the heroes of my world in the hope those in my industry will be inspired to read her books and try to implement psychological safety in their own organisations. If just 100 extra people get the message and catch the psychological safety bug, it would have been worth it.

I am grateful for Duhigg's New York Times article, 'What Google learned from its quest to build the perfect team', which I relied on for the opening paragraphs of Chapter 16, 'The secret life of teams'.

Finally, before I get to thanking those closest to me, I'd like to thank Paul Mason for *Postcapitalism*. It was this book that started my summer adventure back in 2016. That adventure soon took me to Rifkin and Brynjolfsson and McAfee and eventually to Drucker's *Post-Capitalist Society* which, along with all of Drucker's work, have shaped my thinking tremendously. I am grateful to these writers, and especially Mason, whom I swapped notes with as he completed *Clear Bright Future*, for my economic education. I enjoyed moving from their works to those of people like Robert Gordon and his *The Rise and Fall of American Growth*. I am still a better participant in economics than I am a student of it. However, I take great strength from the simple fact, whenever I worry that I don't know what I am talking about, that almost all economists don't know what they're talking about either. Since none of us know what we're talking about, the facts matter less than the conversation. I hope in Part 5 of this book I have added to the conversation.

Now to those closer to me.

First of all, thanks to the team at The Right Book Company. To Beverley Glick, my editor, for coming with me on the journey and for shaping the contours of the book and my skills as a writer. To Andrew Chapman, copy editor extraordinaire. I promised myself that I would not let Andrew edit anything that I could not edit myself. One thing that stumped me was the thorny issue of keeping the reader with me in a book that spans three centuries. I had built scaffolding on which to build the book but once I finished the damn thing I did not know how to dismantle it. That fell to Andrew who helped me take it down and in doing so deal with the earlier, over-summarised versions. The book is better for it and that's why the reader gets way more narrative and way less exposition.

To my younger brother, Joel, who was the sort of reader I intended for this book. Thanks for going over the manuscript and giving me timely and thoughtful feedback. Thanks too for the constant encouragement and curiosity.

In the summer of 2024, I found myself in Paris for KubeCon with Ian Crosby and Ian Miell. They teased me relentlessly when I told them that I was writing about the cloud and had decided the starting point should be Galvani's frogs' legs. Their curiosity eventually got the better of them and they became the first reviewers of this book. Their patient and thoughtful feedback made this manuscript stronger. Thanks lads.

To the rest of my colleagues at Container Solutions, past and present, who for more than a decade have listened, often with rapture, to the stories I love to tell and who encouraged me to collect them all in one place, thanks.

Our journey reflects the story of the cloud itself. In the beginning we were excited to speed up knowing we could scale servers up and down without having to buy them. We knew that would accelerate the software development cycle, which is just another way of saying we knew it would accelerate our creativity and in turn that would allow us to shape our products with end users and their impossible-to-predict behaviour. Soon after, our customers caught the bug and we were delighted to help them. However, as larger enterprises became infected, the ones so large that they had their own centres of gravity, the things we took for granted, psychological safety and the management of systems and those who build them, were brought sharply into focus. What was always true was still true: companies that wanted to succeed with computers had to succeed with management, too. Later, our marvel at Software as a Service quickly changed to concern as the societal implications of our work came into focus. Throughout it all, we did all we could to help our customers, educate ourselves and, whenever the chance came our way, to educate those outside of our industry, too. In an industry that on the one hand is full of the heroes like Licklider and Taylor but on the other is also full of bullies and parasites, we did everything we could to be the good guys, everything we could to counter the technological determinism that haunts the industry but that we know to be bullshit. You all believed in me and I believed in all of you. For this, I am eternally grateful. Failing forward with the team at Container Solutions was and still is the greatest adventure of my life.

Finally, to my family, who put up with me falling asleep at teatime because I was up every Saturday and Sunday in the wee hours, chipping away at this book for more weekends than I care to share. Thank you. My obsession with computers was finally displaced when you all started to arrive 13 years ago. Since then, and forever more, my day will start and end with you, my Alphas and my Omegas, the infinite loop that I never want to break out of.

Bibliography

Allen, P (2012) *Idea Man*. Penguin.
Amdahl, K (2019) *There are No Electrons: Electronics for Earthlings*. Clearwater Publishing.
Anderson, C (2009) *The Longer Long Tail*. Random House Business.
Archives IT (2024) 'LEO – The world's first "Electronic Office"'. URL: archivesit.org.uk/leo
Baran, P (1962) 'On distributed communications networks'. RAND Corporation. URL: rand.org/content/dam/rand/pubs/papers/2005/P2626.pdf
Baran, P (2002) 'The beginnings of packet switching: Some underlying concepts'. *IEEE Communications Magazine*, July. URL ieeexplore.ieee.org/document/1018006
Baran, P (2024) 'Paul Baran and the origins of the internet'. RAND Corporation. URL: rand.org/about/history/baran
Barrs, P (2012) *Erotic Engine*. Doubleday Canada.
Beard, C A (1927) 'Time, technology, and the creative spirit in political science'. *The American Political Science Review* 21(1).
Berlin, L (2005) *The Man Behind the Microchip*. Oxford University Press.
Berlin, L (2017) *Troublemakers*. Simon & Schuster.
Bersin, J & Chamorro-Premuzic, T (2019) 'Hire leaders for what they can do, not what they have done'. *Harvard Business Review*, 27 August. URL: hbr.org/2019/08/hire-leaders-for-what-they-can-do-not-what-they-have-done
Bezos, J P (2021) *Invent and Wander*. Harvard Business Review Press.
Bhattacharya, A (2021) *The Man from the Future*. Penguin Books.

Bird, K & Sherwin, M J (2005) *American Prometheus*. Penguin Random House.

Black, B (2009) 'EC2 origins'. Benjamin Black, 25 January. URL: blog.b3k.us/2009/01/25/ec2-origins.html

Brand, S (1972) 'Spacewar, fantastic life and death among computer bums'. *Rolling Stone*, 7 December.

Brew, S (2019) 'How Fantasia's marketing made a half billion dollars on VHS'. *Den of Geek*, 13 November. URL: denofgeek.com/movies/how-fantasias-marketing-made-a-half-billion-dollars-on-vhs

Brockmeier, E K (2021) '75th anniversary of the electronic numerical integrator and computer (ENIAC)'. The University of Pennsylvania Almanac, 6 April. URL: almanac.upenn.edu/articles/75th-anniversary-of-the-electronic-numerical-integrator-and-computer-eniac

Brown, B L (2020) 'The Bell versus Gray telephone dispute: Resolving a 144-year-old controversy'. *IEEE*, 27 October. URL: ieeexplore.ieee.org/document/9241496

Brown, T (2024) 'Child poverty: Statistics, causes and the UK's policy response'. House of Lords, 23 April. URL: lordslibrary.parliament.uk/child-poverty-statistics-causes-and-the-uks-policy-response

Bryar, C & Carr, B (2021) *Working Backwards*. St Martin's Press.

Brynjolfsson, E & McAfee, A (2012) *Race Against the Machine*. Digital Frontier Press.

Brynjolfsson, E & McAfee, A (2014) *The Second Machine Age*. Norton.

Buxton-Cope, T (2020) *Who the Hell is B.F. Skinner?* Bowden and Brazil.

Cadwalladr, C (2017) 'Revealed: How US billionaire helped to back Brexit'. *The Guardian*, 26 February. URL: theguardian.com/politics/2017/feb/26/us-billionaire-mercer-helped-back-brexit

Campbell, J (1995) *Myths to Live By*. Souvenir.

Canal, A (2024) 'Why Disney's Bob Iger called Netflix "the gold standard" in streaming'. Yahoo! Finance, 7 May. URL: finance.yahoo.com/news/why-disneys-bob-iger-called-netflix-the-gold-standard-in-streaming-151130429.html

Cardwell, D (1994) *The Fontana History of Technology*. Fontana Press.

CERN (2024) 'A short history of the Web'. URL: home.cern/science/computing/birth-web/short-history-web

Ceruzzi, P E (2002) *A History of Modern Computing*, 2nd edition. The MIT Press.

Ceruzzi, P E (2012) *Computing: A concise history*. The MIT Press.

Chamorro-Premuzic, T (2013) 'Why do so many incompetent men become leaders?'. *Harvard Business Review*, 22 August. URL: hbr.org/2013/08/why-do-so-many-incompetent-men

Chamorro-Premuzic, T (2019) *Why Do So Many Incompetent Men Become Leaders?* Harvard Business Review Press.

Cheung, D & Brach, E (2020) *Conquering the Electron*. Lyon Press.

Christakis, N A (2019) 'How AI will rewire us'. *The Atlantic*, April. URL: theatlantic.com/magazine/archive/2019/04/robots-human-relationships/583204

Clark, L (2015) 'DeepMind's AI is an Atari gaming pro now'. *Wired*, 25 February. URL: wired.com/story/google-deepmind-atari

Clarkson, W (2007) *Quentin Tarantino*. John Blake Publishing.

Clifford, C (2018) 'Jeff Bezos says this is how he plans to spend the bulk of his fortune'. CNBC, 30 April. URL: cnbc.com/2018/04/30/jeff-bezos-says-this-is-how-he-plans-to-spend-the-bulk-of-his-fortune.html

Cohen, D (1997) *Carl Rogers*. Constable and Company.

Colquhoun, M (2021) 'Hull and the bomb'. Xenogothic, 2 February. URL: xenogothic.com/2021/02/25/hull-and-the-bomb

Conley, C (2017) *Peak*. J Wiley & Sons.

Courtney, H, Kirkland, J & Viguerie, P (1997) 'Strategy under uncertainty'. *Harvard Business Review* November/December. URL: hbr.org/1997/11/strategy-under-uncertainty

Cox, D, Navarro-Rivera, J & Jones, R P (2013) 'In search of libertarians in America'. 29 October. URL: prri.org/research/2013-american-values-survey

Creighton, T E (1992) *Protein Folding*. W H Freeman and Co.

de Brunner, J T P & Munck, E (1969) '"Levinthal's Paradox" in Mossbauer spectroscopy in biological systems'. University of Illinois Press. URL: web.archive.org/web/20110523080407/http://www-miller.ch.cam.ac.uk/levinthal/levinthal.html

Dixon, N F (1994) *On the Psychology of Military Incompetence*. Pimlico.

Doerr, J (2018) *Measure What Matters*. Penguin Random House.

Drucker, P (1993) *Post-Capitalist Society*. Routledge.

Duhigg, C (2016) 'What Google learned from its quest to build the perfect team'. *New York Times*, 25 February. URL: nytimes.com/2016/02/28/magazine/what-google-learned-from-its-quest-to-build-the-perfect-team.html

Dyson, G (2012) *Turing's Cathedral*. Penguin Books.

Edmondson, A (2018) *The Fearless Organization*. Wiley & Sons.

Edmondson, A (2024) *Right Kind of Wrong*. Penguin.

Eindhoven University of Technology (2018) 'New AI method increases the power of artificial neural networks'. URL: phys.org/news/2018-06-ai-method-power-artificial-neural.html

Ellingrud, K, Sanghvi, S et al (2023) 'Generative AI and the future of work in America'. McKinsey and Co, 26 July. URL: mckinsey.com/mgi/our-research/generative-ai-and-the-future-of-work-in-america

Encyclopaedia Britannica (2024) 'Elisha Gray'. 29 July. URL: britannica.com/biography/Elisha-Gray

Evans, C (1979) *The Mighty Micro: The impact of the computer revolution*. Victor Gollancz Ltd.

Farrington G (1996) 'ENIAC: The birth of the information age'. *Popular Science*, March.

Faulkner, W (1930) *As I Lay Dying*. Jonathan Cape.

Ferry, G (2003) *A Computer Called LEO*. Fourth Estate.

Garfinkel, S (2011) 'The cloud imperative'. *MIT Technology Review*, 3 October. URL: technologyreview.com/2011/10/03/190237/the-cloud-imperative

Garvin, D A, Edmondson A & Gino, F (2008) 'Is yours a learning organization?' *Harvard Business Review*, March. URL: hbr.org/2008/03/is-yours-a-learning-organization

Gertner, J (2012) *The Idea Factory: Bell Labs and the great age of American innovation*. Penguin Books.

Goldman, S (2013) *Science in the 20th Century: A social-intellectual survey*. The Great Courses.

Gordon, R J (2016) *The Rise and Fall of American Growth*. Princeton University Press.

Graetz, G & Michaels, G (2017) 'Is modern technology responsible for jobless recoveries?' *American Economic Review*, May.

Gray, C (2006) *Reluctant Genius*. Arcade Publishing.
Greenberg, J M (2008) *From Betamax to Blockbuster*. MIT Press.
Griffith, E (2022) 'Haters gonna hate: The most-despised tech brands'. *PCMag*, 15 January. URL: uk.pcmag.com/news/135896/haters-gonna-hate-the-most-despised-tech-brand
Grim, P (2012) *Philosophy of Mind*. The Great Courses.
Grove, A (1995) *High Output Management*. Vintage.
Grove, A (1996) *Only the Paranoid Survive*. Profile Books.
Grove, A (2001) *Swimming Across*. Warner Books.
Gulati, R, Ciechanover, A & Huizinga, (2019) 'Netflix: A creative approach to culture and agility'. Harvard Business School, 23 September.
Hafner, K & Lyon, M (1996) *Where Wizards Stay Up Late*. Simon & Schuster.
Haigh, T & Ceruzzi, P E (2021) *A New History of Modern Computing*. The MIT Press.
Halliday, J (2024) 'More than one in three children in poverty as UK deprivation hits record high'. *Guardian*, 19 November. URL: theguardian.com/society/2024/nov/18/more-than-one-in-three-uk-children-poverty-deprivation-record-high
Hanson, D (1982) *The New Alchemists*. Avon.
Harris, K (2020) 'Forty years of falling manufacturing employment'. US Bureau of Labor Statistics, 20 November. URL: bls.gov/opub/btn/volume-9/forty-years-of-falling-manufacturing-employment.htm
Hastings, R & Meyer, E (2020) *No Rules Rules*. Penguin Random House.
Herbert, D (2014) *Videoland*. University of California Press.
Hersey, J (2001) *Hiroshima*. Penguin Classics.
Hiltzik, M A (2000) *Dealers of Lightning*. Harper Business.
History of Computer Communications (2021) 'The history of computer communications'. URL: historyofcomputercommunications.info
Hopkins, C (2023) 'The history of Amazon and its rise to success'. *Michigan Journal of Economics*, 1 May. URL: sites.lsa.umich.edu/mje/2023/05/01/the-history-of-amazon-and-its-rise-to-success
Hughes, T P (2004) *American Genesis*. The University of Chicago Press.
Institute for Public Policy Research (2024) 'Up to 8 million UK jobs at risk from AI unless government acts, finds IPPR'. 27 March. URL:

ippr.org/media-office/up-to-8-million-uk-jobs-at-risk-from-ai-unless-government-acts-finds-ippr

Intel (nd), 'Explore Intel's History'. URL: timeline.intel.com

Israel, P (1998) *Edison: A life of invention*. John Wiley & Sons, Inc.

Izrailevsky, Y, Vlaovic, S & Meshenberg, R (2016) 'Completing the Netflix cloud migration'. Netflix, 12 February. URL: about.netflix.com/en/news/completing-the-netflix-cloud-migration

Jackson, A (2023) 'These are the 7 most hated brands in America—Elon Musk's Twitter is No. 4'. CNBC, 1 June. URL: cnbc.com/2023/06/01/most-hated-brands-in-america-trump-organization-ftx-fox-corporation.html

Jackson, T (1997) *Inside Intel*. HarperCollins.

Jaimovich, N & Siu, H (2012) 'The trend is the cycle: Job polarization and jobless recoveries'. *Review of Economics and Statistics* 102 (18334), August.

Jarow, O (2023) 'We cut child poverty to historic lows, then let it rebound faster than ever before'. *Vox*, 21 September. URL: vox.com/future-perfect/2023/9/21/23882353/child-poverty-expanded-child-tax-credit-census-welfare-inflation-economy-data

Johnson, P (2022) 'How Hull would have fought a nuclear war'. *Hull Daily Mail,* 3 July. URL: hulldailymail.co.uk/news/history/how-hull-fought-nuclear-war-7275384

Jorgensen, T J (2021) *Spark: The life of electricity and the electricity of life*. Princeton University Press.

Journal Staff Report (1995) 'US Atomic Agency ex-official, 80, dies'. *Albuquerque Journal* 16 November. URL: newspapers.com/article/albuquerque-journal-lawrence-preston-gis/35782841

Judis, J B (2016) *The Populist Explosion*. Columbia Global Reports.

Jumper, J, Evans, R et al (2021) 'Highly accurate protein structure prediction with AlphaFold'. *Nature,* 15 July. URL: nature.com/articles/s41586-021-03819-2

Kaplan, F (2017) 'How an agency of oddballs transformed modern war and modern life'. *New York Times*, 30 June. URL: nytimes.com/2017/06/30/books/review/imagineers-of-war-untold-history-of-darpa-sharon-weinberger.html

Kasparov, Garry (2017) 'Don't fear intelligent machines. Work with them'. TED, April. URL: ted.com/talks/garry_kasparov_don_t_fear_ intelligent_machines_work_with_them

Kennedy Jr, T R (1954) 'Electronic computer flashes answers, may speed engineering'. *New York Times*, 15 February. URL: nytimes.com/1946/02/15/archives/electronic-computer-flashes-answers-may-speed-engineering-new.html

Keynes, J M (2009) *Essays in Persuasion*. CreateSpace.

Kilby, J (2000) 'Turning potential into realities: The invention of the integrated circuit: Nobel Lecture'. The Nobel Prize. URL: nobelprize.org/prizes/physics/2000/kilby/lecture

King, R (2011) 'Larry Page: Google+ is just beginning, will become "automagical"'. ZDNET, 13 October. URL: zdnet.com/article/larry-page-google-is-just-beginning-will-become-automagical

Kirby, J & Stewart, T A (2007) 'The institutional yes'. *Harvard Business Review*, October. URL: hbr.org/2007/10/the-institutional-yes

Kohs, G (2017) *AlphaGo*. Moxie Pictures, Reel As Dirt, 29 September. URL: youtube.com/watch?v=WXuK6gekU1Y

Koutavas, A, Year, C et al (2024) 'What could 2023 child poverty rates have looked like had an expanded Child Tax Credit had still been in place? A poverty reduction analysis of the 2023 American Family Act'. *Poverty and Social Policy Brief* 8(3), Columbia School of Social Work, 10 September. URL: povertycenter.columbia.edu/publication/what-2023-child-poverty-rates-could-have-looked-like

Krugman, P (2023) 'The internet was an economic disappointment'. *New York Times*, 4 April. URL: nytimes.com/2023/04/04/opinion/internet-economy.html

Landes, D (2003) *The Unbound Prometheus*, 2nd edition. Cambridge University Press.

Lardner, J (1987) *Fast Forward*. Norton.

Lay, K & Okiror, S (2024) 'I am happy to see how my baby is bouncing': The AI transforming pregnancy scans in Africa'. *The Guardian*, 12 July. URL: theguardian.com/global-development/article/2024/jul/12/i-am-happy-to-see-how-my-baby-is-bouncing-the-ai-transforming-pregnancy-scans-in-africa

Lean, T (2016) *Electronic Dreams*. Bloomsbury Sigma.

Leo Computer Society (2024) 'Leo Computer Society'. URL: leo-computers.org.uk

Levy, S (2010) *Hackers*. O'Reilly.

Licklider, J C R (1960) 'Man-computer symbiosis'. IRE Transactions on Human Factors in Electronics, March. URL: groups.csail.mit.edu/medg/people/psz/Licklider.html

Licklider, J C R (1963) 'Memorandum for members and affiliates of the Intergalactic Computer Network'. ARPA. URL: thekurzweillibrary.com/memorandum-for-members-and-affiliates-of-the-intergalactic-computer-network

Litman, J (2006) 'The story of Sony v Universal Studios: Mary Poppins meets the Boston Strangler'. University of Michigan. URL: repository.law.umich.edu/cgi/viewcontent.cgi?article=1214

Locke, E A & Latham, G P (2013) *New Developments in Goal Setting and Task Performance*. Routledge.

Malone, M (2014) *The Intel Trinity*. HarperCollins.

Manes, S & Andrews, P (1993) *Gates*. Simon & Schuster.

Markoff, J (2017) 'Robert Taylor, innovator who shaped modern computing, dies at 85'. *New York Times*, 14 April. URL: nytimes.com/2017/04/14/technology/robert-taylor-innovator-who-shaped-modern-computing-dies-at-85.html

Marsh, A (2021) 'Inside the Third Reich's radio'. *IEEE Spectrum*, March. URL: spectrum.ieee.org/inside-the-third-reichs-radio

Marshall, D (2004) *Bill Gates*. Blackbirch Press.

Maslow, A (1943) 'A theory of human motivation'. Originally published in *Psychological Review*, 50, 370-396. URL: psychclassics.yorku.ca/Maslow/motivation.htm

Maslow, A (1975) *The Farther Reaches of Human Nature*. Viking Press.

Maslow, A (1997) *Motivation and Personality*, third edition. Pearson.

Maslow, A (1998) *Maslow on Management*. J Wiley & Sons.

Maslow, A (1998b) *Toward a Psychology of Being*. J Wiley & Sons.

Mason, P (2016) *PostCapitalism*. Penguin.

Mason, P (2019) *Clear Bright Future*. Allen Lane.

McComas, A J (2011) *Galvani's Spark: The story of the nerve impulse*. Oxford University Press.

McCord, P (2014) 'How Netflix reinvented HR'. *Harvard Business Review*, January. URL: hbr.org/2014/01/how-netflix-reinvented-hr

McCord, P (2017) *Powerful*. Silicon Guild.

McManus, (2023) 'Libertarians weren't always apologists for the rich and powerful'. *The Jacobin*, 4 September. URL: jacobin.com/2023/04/libertarians-right-left-capitalism-socialism-mises-rand

Meier, A C (2018) 'An affordable radio brought Nazi propaganda home'. JSTOR, 30 August. URL: daily.jstor.org/an-affordable-radio-brought-nazi-propaganda-home

Merchant, B (2023) *Blood in the Machine*. Little, Brown and Company.

Miller, C (2022) *Chip War*. Simon & Schuster.

Mitchell, M (2019) *Artificial Intelligence*. Penguin Random House.

Mokry, J (2009) *The Enlightened Economy*. Penguins Books.

Morita, A (1986) *Made in Japan*. E P Dutton.

Morley, N (2021) *Radio Hitler*. Amberley Publishing.

Morris, E (2019) *Edison*. Random House.

Motukuri, K (2022) 'Ambient computing is at the heart of the frictionless economy'. *Forbes*, 21 November. URL: forbes.com/councils/forbestechcouncil/2022/11/21/ambient-computing-is-at-the-heart-of-the-frictionless-economy

National Human Genome Research Institute (2024) 'The cost of sequencing a human genome'. URL: genome.gov/about-genomics/fact-sheets/Sequencing-Human-Genome-cost

National Museum Scotland (2024) 'Did Alexander Graham Bell invent the telephone?' URL: nms.ac.uk/discover-catalogue/did-alexander-graham-bell-invent-the-telephone

Netflix (2024) 'Netflix culture – the best work of our lives'. Netflix. URL: jobs.netflix.com/culture.

Noble, D F (1979) *America by Design*. Oxford University Press.

Nolan, C (2023) *Oppenheimer: The official screenplay*. Faber & Faber.

O'Mara, M (2018) 'Silicon Valley can't escape the business of war'. *New York Times*, 23 October. URL: nytimes.com/2018/10/26/opinion/amazon-bezos-pentagon-hq2.html

O'Mara, M (2020) *The Code*. Penguin Random House.

O'Regan (2008) *A Brief History of Computing*. Springer.

Olson, P (2024) *Supremacy*. St Martin's Press.

Oppenheimer, R J (2023) 'Oppenheimer replies'. *Bulletin of the Atomic Scientists*, 17 July. URL: thebulletin.org/premium/2023-07/oppenheimer-replies

Ortiz, S (2022) 'What is ambient computing? Everything you need to know about the rise of invisible tech'. ZDNET, 13 September. URL: zdnet.com/article/what-is-ambient-computing-everything-you-need-to-know-about-the-rise-of-invisible-tech

Payne, A (2021) *Built to Fail*. Lioncrest.

Penn Engineering (2017) 'Celebrating Penn Engineering history: ENIAC'. URL: seas.upenn.edu/about/history-heritage/eniac

Pierce, D (2022) 'Pixel by Pixel: how Google is trying to focus and ship the future'. *The Verge*, 11 May. URL: theverge.com/23065820/google-io-ambient-computing-pixel-android-phones-watches-software

Porter, B & Machery, E (2024) 'AI-generated poetry is indistinguishable from human-written poetry and is rated more favourably'. *Sci Rep* 14, 26133 URL: doi.org/10.1038/s41598-024-76900-1

Randolph, M (2021) *That Will Never Work*. Endeavour.

Reid, T R (1985) *The Chip*. Random House.

Reuter, D (2021) 'Austin Powers' super villain Dr Evil was trending in connection with Jeff Bezos' trip to space, and these photos show why'. *Business Insider* 20 July. URL: businessinsider.com/jeff-bezos-compared-to-austin-powers-villain-dr-evil-2021-7

Rhodes, R (2012) *The Making of the Atomic Bomb*. Simon & Schuster.

Rifkin, J (2014) *Zero Marginal Cost Society*. Griffin.

Robertson, R (2020) *The Enlightenment*. Penguin Random House.

Rose, C (2009) 'Jeff Bezos'. Charlie Rose, 26 February. URL: charlierose.com/videos/22164

Rose, C (2014) 'Steve Ballmer'. Charlie Rose, 21 October. URL: charlierose.com/videos/28129

Rosen, R J (2012) 'Time and space has been completely annihilated'. *The Atlantic*, 14 February. URL: theatlantic.com/technology/archive/2012/02/time-and-space-has-been-completely-annihilated/253103

Rossman, J (2021) *The Amazon Way*. Clyde Hill Publishing.

Ryan, (2010) *A History of the Internet and the Digital Future*. Reaktion Books.

Schewe, P F (2007) *The Grid*. Joseph Henry Press.

Schwartz, M (2017) 'Facebook failed to protect 30 million users from having their data harvested by Trump campaign affiliate'.

The Intercept, 30 March. URL: theintercept.com/2017/03/30/facebook-failed-to-protect-30-million-users-from-having-their-data-harvested-by-trump-campaign-affiliate

Science and Industry Museum (2019) 'Programming patterns: the story of the Jacquard loom'. Science and Industry Museum, 25 June. URL: scienceandindustrymuseum.org.uk/objects-and-stories/jacquard-loom

Shelley, M (2012) *Frankenstein*. Penguin Classics.

Shih, W C, Kaufman S P & Spinola, D (2009) 'Netflix'. Harvard Business School, 27 April. URL: hbs.edu/faculty/Pages/item.aspx?num=34596

Shulman, S (2008) *The Telephone Gambit: Chasing Alexander Graham Bell's secret*. Norton.

Shurkin, J N (2006) *Broken Genius: The rise and fall of William Shockley creator of the electronic age*. MacMillan.

Silverman, K (2003) *Lightning Man*. De Capo Press.

Smil, V (2005) *Creating the Twentieth Century*. Oxford University Press.

Smil, V (2015) 'The Miraculous 1880s'. *IEEE Spectrum*, 10 July. URL: vaclavsmil.com/wp-content/uploads/2024/10/7.1880s.pdf

Smil, V (2017) *Energy: A beginner's guide*. Oneworld Publications.

Smith, D K & Alexander, R C (1999) *Fumbling the Future: How Xerox invented, then ignored, the first personal computer*. toExcel.

Sperling, N (2024) 'Responsibility over freedom: How Netflix's culture has changed'. *New York Times*, 24 June. URL: nytimes.com/2024/06/24/business/media/netflix-corporate-culture.html

Standage, T (1998) *The Victorian Internet*. Phoenix.

Stoker, B (2003) *Dracula*. Penguin Classics.

Stone, B (2014) *The Everything Store*. Penguin Random House.

Stone, B (2021) *Amazon Unbound*. Simon & Schuster.

Strother, R (2007) *Bill Gates*. ABDO Publishing Company.

Suleyman, M & Bhaskar, M (2023) *The Coming Wave*. Penguin Random House.

Tedlow, R S (2006) *Andy Grove*. Penguin.

The Science Museum (2018) 'Ahoy! Alexander Graham Bell and the first telephone call'. The Science Museum, 19 October. URL: sciencemuseum.org.uk/objects-and-stories/ahoy-alexander-graham-bell-and-first-telephone-call

Tiku, N (2018) 'Amazon's Jeff Bezos says tech companies should work with the Pentagon'. *Wired*, 15 October. URL: wired.com/story/amazons-jeff-bezos-says-tech-companies-should-work-with-the-pentagon

Toews, R (2021) 'AlphaFold is the most important achievement in AI – ever'. *Forbes*, 3 October. URL: forbes.com/sites/robtoews/2021/10/03/alphafold-is-the-most-important-achievement-in-ai-ever

Transport for London (2024) 'Learn the knowledge of London'. URL: tfl.gov.uk/info-for/taxis-and-private-hire/licensing/learn-the-knowledge-of-london

Umoh, R (2017) 'Jeff Bezos says he learned this critical business skill while visiting his grandfather as a child'. CNBC, 17 November. URL: cnbc.com/2017/11/17/jeff-bezos-learned-this-critical-business-skill-from-his-grandfather.html

United Nations (2019) 'New report calls for urgent action to avert antimicrobial resistance crisis'. United Nations, 29 April. URL: who.int/news/item/29-04-2019-new-report-calls-for-urgent-action-to-avert-antimicrobial-resistance-crisis

Verne, J (1870) *Twenty Thousand Leagues under the Sea.*

Waldrop, M M (2018) *The Dream Machine*. Stripe Press.

Weinberger, S (2017) *The Imagineers of War*. Vintage Books.

Wills, I (2019) *Thomas Edison: Success and innovation through failure*. Springer.

Winterson, J (2022) *12 Bytes*. Vintage.

Wolfe, T (2018) 'The tinkerings of Robert Noyce'. *Esquire*, 15 May. URL: esquire.com/news-politics/a12149389/robert-noyce-tom-wolfe

Wong, J C (2019) 'Facebook to be fined $5bn for Cambridge Analytica privacy violations – reports. *The Guardian*, 12 July. URL: theguardian.com/technology/2019/jul/12/facebook-fine-ftc-privacy-violations

Wu, T (2017) *The Attention Merchants*. Vintage Books.